Functional and Structural Proteins of the Nervous System

ADVANCES IN EXPERIMENTAL MEDICINE AND BIOLOGY

Volume 1
THE RETICULOENDOTHELIAL SYSTEM AND ATHEROSCLEROSIS
Edited by N. R. Di Luzio and R. Paoletti • 1967

Volume 2
PHARMACOLOGY OF HORMONAL POLYPEPTIDES AND PROTEINS
Edited by N. Back, L. Martini, and R. Paoletti • 1968

Volume 3
GERM-FREE BIOLOGY: Experimental and Clinical Aspects
Edited by E. A. Mirand and N. Back • 1969

Volume 4
DRUGS AFFECTING LIPID METABOLISM
Edited by W. L. Holmes, L. A. Carlson, and R. Paoletti • 1969

Volume 5
LYMPHATIC TISSUE AND GERMINAL CENTERS IN IMMUNE RESPONSE
Edited by L. Fiore-Donati and M. G. Hanna, Jr. • 1969

Volume 6
RED CELL METABOLISM AND FUNCTION
Edited by George J. Brewer • 1970

Volume 7
SURFACE CHEMISTRY OF BIOLOGICAL SYSTEMS
Edited by Martin Blank • 1970

Volume 8
BRADYKININ AND RELATED KININS: Cardiovascular, Biochemical, and Neural Actions
Edited by F. Sicuteri, M. Rocha e Silva, and N. Back • 1970

Volume 9
SHOCK: Biochemical, Pharmacological, and Clinical Aspects
Edited by A. Bertelli and N. Back • 1970

Volume 10
THE HUMAN TESTIS
Edited by E. Rosemberg and C. A. Paulsen • 1970

Volume 11
MUSCLE METABOLISM DURING EXERCISE
Edited by B. Pernow and B. Saltin • 1971

Volume 12
MORPHOLOGICAL AND FUNCTIONAL ASPECTS OF IMMUNITY
Edited by K. Lindahl-Kiessling, G. Alm, and M. G. Hanna, Jr. • 1971

Volume 13
CHEMISTRY AND BRAIN DEVELOPMENT
Edited by R. Paoletti and A. N. Davison • 1971

Volume 14
MEMBRANE-BOUND ENZYMES
Edited by G. Porcellati and F. di Jeso • 1971

Volume 15
THE RETICULOENDOTHELIAL SYSTEM AND IMMUNE PHENOMENA
Edited by N. R. Di Luzio and K. Flemming • 1971

Volume 16A
THE ARTERY AND THE PROCESS OF ARTERIOSCLEROSIS: Pathogenesis
Edited by Stewart Wolf • 1971

Volume 16B
THE ARTERY AND THE PROCESS OF ARTERIOSCLEROSIS: Measurement and Modification
Edited by Stewart Wolf • 1971

Volume 17
CONTROL OF RENIN SECRETION
Edited by Tatiana A. Assaykeen • 1972

Volume 18
THE DYNAMICS OF MERISTEM CELL POPULATIONS
Edited by Morton W. Miller and Charles C. Kuehnert • 1972

Volume 19
SPHINGOLIPIDS, SPHINGOLIPIDOSES AND ALLIED DISORDERS
Edited by Bruno W. Volk and Stanley M. Aronson • 1972

Volume 20
DRUG ABUSE: Nonmedical Use of Dependence-Producing Drugs
Edited by Simon Btesh • 1972

Volume 21
VASOPEPTIDES: Chemistry, Pharmacology, and Pathophysiology
Edited by N. Back and F. Sicuteri • 1972

Volume 22
COMPARATIVE PATHOPHYSIOLOGY OF CIRCULATORY DISTURBANCES
Edited by Colin M. Bloor • 1972

Volume 23
THE FUNDAMENTAL MECHANISMS OF SHOCK
Edited by Lerner B. Hinshaw and Barbara G. Cox • 1972

Volume 24
THE VISUAL SYSTEM: Neurophysiology, Biophysics, and Their Clinical Applications
Edited by G. B. Arden • 1972

Volume 25
GLYCOLIPIDS, GLYCOPROTEINS, AND MUCOPOLYSACCHARIDES OF THE NERVOUS SYSTEM
Edited by Vittorio Zambotti, Guido Tettamanti, and Mariagrazia Arrigoni • 1972

Volume 26
PHARMACOLOGICAL CONTROL OF LIPID METABOLISM
Edited by William L. Holmes, Rodolfo Paoletti, and David Kritchevsky • 1972

Volume 27
DRUGS AND FETAL DEVELOPMENT
Edited by Marcus A. Klingberg, Armand Abramovici, and Juan Chemke • 1972

Volume 28
HEMOGLOBIN AND RED CELL STRUCTURE AND FUNCTION
Edited by George J. Brewer • 1972

Volume 29
MICROENVIRONMENTAL ASPECTS OF IMMUNITY
Edited by Branislav D. Janković and Katarina Isaković • 1972

Volume 30
HUMAN DEVELOPMENT AND THE THYROID GLAND: Relation to Endemic Cretinism
Edited by J. B. Stanbury and R. L. Kroc • 1972

Volume 31
IMMUNITY IN VIRAL AND RICKETTSIAL DISEASES
Edited by A. Kohn and M. A. Klingberg • 1972

Volume 32
FUNCTIONAL AND STRUCTURAL PROTEINS OF THE NERVOUS SYSTEM
Edited by A. N. Davison, P. Mandel, and I. G. Morgan • 1972

Functional and Structural Proteins of the Nervous System

Proceedings of Two Symposia on Proteins of the Nervous System and Myelin Proteins Held as Part of the Third Meeting of the International Society of Neurochemistry in Budapest, Hungary, in July 1971

Edited by

A. N. Davison

Department of Neurochemistry
Institute of Neurology
London, England

and

P. Mandel

and

I. G. Morgan

Centre de Neurochemie du CNRS
Strasbourg, France

PLENUM PRESS • NEW YORK-LONDON • 1972

Library of Congress Catalog Card Number — 72-91937
ISBN 0-306-39032-9

A Division of Plenum Publishing Corporation
227 West 17th Street, New York, N.Y. 10011

United Kingdom edition published by Plenum Press, London
A Division of Plenum Publishing Company
Davis House (4th Floor), 8 Scrubs Lane, Harlesden, London,
NW10 6SE, England

Printed in the United States of America

PREFACE

This book is the proceedings of two symposia, on Proteins of the Nervous System and Myelin Proteins, held as part of the Third Meeting of the International Society of Neurochemistry in Budapest, July 1971. If confirmation of the utility for such a book was needed, it was obtained with the independent publication of a review on Proteins of the Nervous System (1) covering essentially the same ground. This review is an excellent collection of the earlier literature. The present volume summarizes the more recent advances in this field, and in addition covers those areas, glycoproteins and peptide hydrolases, which were not dealt with by Shooter and Einstein (1).

We are most grateful to Miss V. Troeger for typing the manuscripts and to Mr. A. Landmann for assistance with photography.

A.N. DAVISON
I.G. MORGAN
P. MANDEL

---

(1) SHOOTER, E.M. and EINSTEIN, E.R., Ann. Rev. Biochem., 40 (1971) 635.

CONTENTS

SECTION 1
NEUROSPECIFIC PROTEINS OF UNKNOWN FUNCTION

SECTION 2
PROTEINS OF FUNCTIONAL IMPORTANCE IN THE NERVOUS SYSTEM

SECTION 3
CEREBRAL GLYCOPROTEINS

SECTION 4
MYELIN PROTEINS

SECTION 1

NEUROSPECIFIC PROTEINS OF UNKNOWN FUNCTION

INTRODUCTORY REMARKS

P. MANDEL

Centre de Neurochimie du CNRS, 67-Strasbourg (France)

Nervous tissue proteins can be classified into five groups :

1) Proteins which have similar structures in most mammalian tissues, as is the case for most of the enzymes of basic metabolic pathways.

2) Enzymes which exist in other tissues, but which are present in different forms in the nervous system. One example of this is tyrosine α-ketoglutarate amino transferase (1, 2). This enzyme,which plays a fundamental role in tyrosine degradation probably has an important regulatory role in the nervous system since it is one of the enzymes, the other being tyrosine hydroxylase, which determines whether tyrosine is degraded or used for catecholamine biosynthesis.

Brain tyrosine transaminase is primarily localized on the mitochondrial membrane, particularly in nerve-ending mitochondria. By contrast liver tyrosine transaminase is soluble. This attachment to the membrane may be related to the high amount of proline in the brain enzyme compared to that in liver. In addition, in brain, where the uptake of pyridoxal phosphate is not as easy as in liver, the coenzyme is strongly bound to the enzyme whereas in liver it is loosely bound. The liver enzyme has a α_4 structure, the brain enzyme $\alpha_2\beta_2$. Both enzymes are competitively inhibited by noradrenaline. The properties of the two enzymes are compared in the table.

3) Enzymes involved in the biosynthesis of compounds existing only in a limited number of tissues or specific to the central nervous system,like galactolipids.

Comparison of Liver TAT and Brain TAT

	Liver TAT	Brain TAT
Localization	Cytosol	Mitochondria
M.W.	115,000	100,000
Subunits	4 (type α_4)	4 (type $\alpha_2\beta_2$)
Pyridoxal-5'-phosphate (PLP)	4 mole	2 mole
PLP binding	Weak	Strong
Amino acids :		
- Proline	6	79
- Basic	117	136
- Acid	210	169

4) Proteins which belong to structures specific to the nervous system such as the proteins of the myelin sheath and synaptic membranes, etc.

5) Neuronal or glial specific proteins or enzymes of unknown function like the proteins 14.3.2 and S-100, and the enzyme 2',3'-cyclic AMP-3'-phosphohydrolase.

Concerning the S-100 protein, with the present data it seems possible to make some speculation. This highly acidic protein becomes hydrophobic in the presence of Ca^{++}. Thus it seems likely that during the post-depolarization flux of cations into the synaptic cleft and then into the glial cells, the S-100 could fix cations to its negative changes. In the presence of Ca^{++}, these cations could be sequestered inside a predominately hydrophobic structure, avoiding changes in osmotic pressure with the glial cells. Removal of Ca^{++} could provoke release of these sequestered cations when the protein reverts to a hydrophilic form.

In this symposium we will deal successively with proteins of functional significance in the central nervous system regardless of their localization and then with proteins specific to the nervous system. Finally we will discuss brain, and more specifically synaptic glycoproteins in view of the increasing evidence of the important functions of these molecules.

REFERENCES

1 AUNIS, D., MARK, J. and MANDEL, P., Life Sci., 10 (1971) 617.
2 MARK, J., PUGGE, H. and MANDEL, P., J. Neurochem., 17 (1970) 1393.

CHEMISTRY AND BIOLOGY OF TWO BRAIN-SPECIFIC PROTEINS, S-100 AND 14-3-2

B.W. MOORE

Department of Psychiatry, Barnes and Renard Hospitals
St. Louis, Missouri 63110 (USA)

During development the cells of the nervous system differentiate greatly, leading to the presence in mature brain of many highly specific forms and functions such as the occurrence of the two basic cell types, neurons and glia ; cell processes of neurons and glia ; microtubular systems ; the apparatus at nerve endings including metabolic systems for transmitter synthesis and degradation, vesicles, and specialized membranes ; propagation of action potentials ; axoplasmic flow ; synaptic transmission ; and the establishment of specific connections. The way in which genetic information is expressed during differentiation is mediated through synthesis of specific protein molecules ; therefore there should be proteins which are specific to the nervous system related to specific neural forms and functions. Furthermore, if these forms and functions are truly general, such proteins should be present in nervous systems of all species. Our work has focussed on looking for nervous system-specific, species non-specific proteins.

The general scheme we have used to search for such proteins in brain is as follows :

(1) "Protein maps" are prepared for brain, liver, and other organs by DEAE-cellulose chromatography followed by acrylamide gel electrophoresis. Proteins are chosen which are presumptively nervous system specific and these proteins are purified.

(2) Antiserum is prepared to the pure protein and used in an immunoassay (complement fixation) for the protein.

(3) The protein is tested for specificity to the nervous system and species non-specificity using the highly specific and sensitive immunoassay.

(4) The protein is localized in cell type, and followed during

TABLE I

Proteins Isolated from Beef Brain

	Molecular weights	
	Centrifugation	SDS gel electrophoresis
S-100	21,000	6,300
14-3-2	43,000	54,000
14-3-3	60,000	32,000
15-4-2	38,000	20,000
AG-1	43,500	50,000
TC-1	23,000	25,000
A-P-1	42,000	21,000

development and function using the immunoassay to measure the protein quantitatively.

Seven presumptively nervous system-specific proteins have been purified from beef brain and, of these, two (S-100 and 14-3-2) have been shown to be specific to the nervous system and species non-specific (9,10,11). The estimated molecular weights of the native proteins and of any subunits are shown in Table I. All are fairly small and acidic proteins.

Cellular Localization of S-100 and 14-3-2

The S-100 protein has been shown to be solely in glia (or Schwann cells) and 14-3-2 primarily in neurons by immunofluorescence, by behaviour during Wallerian degeneration of optic nerve (13) or retrograde degeneration of dorsal thalamus (6), by measurements on glioma and neuroblastoma tissue cultures (1, 4), and by measurements in single dissected cells (12).

During development of the chick optic tectum (5) S-100 and 14-3-2 levels rise rapidly in parallel, during the period immediately before and after hatching, only several days after cell division has ceased and the tectum looks histologically mature. In chick spinal cord, the largest rise in S-100 and 14-3-2 occurs at the time when electrical activity and limb movements begin (3). In human brain, the development of S-100 in histologically late-developing areas (frontal cortex) lags behind its development in earlier areas such as pons and cerebellum (15). In the C-6 glioma cell cultures, S-100 is synthesized in large amounts only when the culture becomes confluent and stops growing (1). These facts indicate that S-100 and 14-3-2 probably are connected with highly differentiated functions of nervous system cells, either glia or neurons.

A clue to the functions of S-100 may be indicated by some of its known chemical properties, particularly its reaction with Ca^{++}. Binding of Ca^{++} leads to the appearance of multiple forms of S-100 on acrylamide gel electrophoresis (2,7,8,14). Binding Ca^{++} also causes a conformational change resulting in exposure of one tryptophan, one tyrosine, several phenylalanine, and two cysteine residues (2). This change is antagonized by monovalent cations, the order of effectiveness being $K^{+} > Na^{+} > Li^{+}$.

REFERENCES

1 BENDA, P., LIGHTBODY, J., SATO, G., LEVINE, L. and SWEET, W., Science, 161 (1968) 370.

2 CALISSANO, P., MOORE, B.W. and FRIESEN, A., Biochemistry, 8 (1969) 4318.

3 CICERO, T.J., unpublished.

4 CICERO, T.J. and MOORE, B.W., unpublished.

5 CICERO, T.J., COWAN, W.M. and MOORE, B.W., Brain Res., 24 (1966) 1.

6 CICERO, T.J., COWAN, W.M., MOORE, B.W. and SUNTZEFF, V., Brain Res., 18 (1970) 25.

7 GOMBOS, G., VINCENDON, G., TARDY, J. and MANDEL, P., C.R.Acad. Sci.(Paris) Sér.D., 263 (1966) 1533.

8 HYDEN, H. and McEWEN, B., Próc. Natl. Acad. Sci. US, 55 (1966) 354.

9 MOORE, B.W., Biochem. Biophys. Res. Commun., 19 (1965) 739.

10 MOORE, B.W., in Handbook of Neurochemistry (Ed. LAJTHA, A.) Plenum Press, 1969, vol. 1, p.93.

11 MOORE, B.W. and PEREZ, V.J., in Physiological and Biochemical Aspects of Nervous Integration (Ed. CARLSON, F.D.), Prentice-Hall, Englewood Cliffs, New Jersey, 1968, p.343.

12 PEREZ, V.J. and MOORE, B.W., unpublished.

13 PEREZ, V.J., OLNEY, J.W., CICERO, T.J., MOORE, B.W. and BAHN, B.A., J. Neurochem., 17 (1970) 511.

14 VINCENDON, G., WAKSMAN, A., UYEMURA, K., TARDY, J. and GOMBOS, G., Arch. Biochem. Biophys., 120 (1967) 233.

15 ZUCKERMAN, J.E., HERSCHMANN, H.R. and LEVINE, L., J. Neurochem., 17 (1970) 247.

THE HETEROGENEITY OF THE S-100 PROTEIN FRACTION

G. VINCENDON, J.P. ZANETTA and G. GOMBOS

Centre de Neurochimie du CNRS and Institut de Chimie Biologique, Faculté de Médecine de Strasbourg, Université Louis-Pasteur, 67-Strasbourg (France)

It is commonly accepted that the neurospecific protein S-100 is a single molecular species, on the basis of immunological data, and on the basis of its behaviour in several chromatographic and electrophoretic systems (10) and during analytical ultracentrifugation (15). In fact, the S-100 protein fraction of several animal species migrated as a single band when brain extracts (or the purified protein) were electrophoresed, in the presence of 0.1 mM EDTA, in the Tris-glycine system of Calissano, Moore and Friesen (1). In addition, "S-100 proteins" from the brains of different mammal species, although having somewhat different amino acid compositions (10, 14), appeared to be unusually similar when cross reacting with antiserum to bovine S-100 (10, 13). The precipitin bands neither intersected nor showed "spurs" on a double diffusion test, with the specific antiserum of Levine and Moore (9).

In 1966 (6) we reported on the electrophoretic heterogeneity of bovine S-100, prepared by us or supplied by Dr. Moore. The electrophoretic heterogeneity depended on the reticulation of the polyacrylamide gel, the buffer used and the concentration of EDTA in the buffer. In the system cited above, we observed the one band pattern reported by Calissano *et al.* (1). However S-100 protein electrophoresed in the presence of EDTA (from 0.1 to 2.5 mM) in continuous buffer systems, where glycine was replaced by HCl or acetic acid,resolved into two immunologically similar protein bands (8,13). The precipitin bands neither intersected nor showed "spurs" on a double diffusion test, with the specific antiserum of Levine and Moore (9).

In view of the possibility that the two electrophoretic bands represented two (at least) distinct molecular species, we have adopted the term F-S100 (S-100 protein fraction) and we call the two bands, fast and slow migrating components (FMC and SMC). The two components were present in different, but reproducible, amounts and proportions in the brains of different animal species (12). In a given species, they showed strong, but reproducible, regional variations (4, 5).

Different extraction and electrophoretic conditions were examined on the brain extracts of three species (rat, beef and pig) in which the faster migrating band was present in extremely different proportions (respectively 95, 60 and 18 percent of total F-S100). The species-specific electrophoretic profiles in the Tris-acetic acid-EDTA system, were not modified by 2-mercaptoethanol in the extraction buffer or in the electrophoretic system, by the EDTA (2.5 mM) alone or together with 2-mercaptoethanol in the extraction buffer, by whether the gels were preelectrophoresed or not, and by incubating the brain for up to 4 h at 37° (7). Both components, eluted directly from the gel after electrophoresis of bovine F-S100 in Tris-acetate-EDTA buffer, migrated as a single band when re-electrophoresed (13). Both components had the same sedimentation coefficient as bovine F-S100 (7). They could not be separated by gel filtration on Sephadex G-200 but a partial separation on the basis of charge was obtained by continuous flow paper electrophoresis, DEAE-Sephadex chromatography, and isoelectric focusing (7).

These data indicated that the two band pattern was due neither to *post-mortem* autolysis, nor to a reversible equilibrium between two different states of a unique molecule, provoked either by the oxidation of SH groups or by the action of bivalent cations. Aggregation phenomena and splitting into subunits could be excluded.

A possible explanation for the apparent contradiction between the electrophoretic heterogeneity of F-S100 and the immunological data was that the two bands represented two different irreversible conformational states of one single molecular species.

In fact, Moore, Calissano, Friesen and Rusca (1, 2, 11) demonstrated conformational changes of S-100 in the presence of calcium ions resulting in multiple band patterns on electrophoresis. They also reported that, at high calcium concentrations, S-100 was resolved into two bands (11). In view of the fact that the effect of Ca^{++} appeared to be irreversible after a prolonged incubation, it could not be excluded that S-100, a single molecular species, existed *in vivo* in two Ca^{++} induced irreversible conformational states which migrated as two bands, or that Ca^{++}, contaminating the buffers, artifactually induced a two band electrophoretic profile. The latter hypothesis is however unlikely since identical electrophoretic patterns were obtained for a given species whether brain soluble proteins were extracted in presence or absence of EDTA (or EDTA and mercapto-ethanol), and since electrophoresis was carried out in the presence of high concentrations of EDTA. The interspecific differences, however, cannot be explained simply by a Ca^{++} effect <u>in vivo</u>, since the concentrations of Ca^{++} determined in the brain extracts of rat, pig and beef were practically the same (7). It is possible however that the reported differences in amino acid composition of F-S100 of different species could cause different reactivities towards Ca^{++}, and thus the different electrophoretic profiles.

Purified pig brain S-100 electrophoresed in the presence of calcium (4 mM) separated into two bands which migrated well behind bromophenol blue (7). A densitometric estimation of the relative proportion of the two bands obtained in the presence of 4 mM calcium indicated that they were present in approximately equal amounts (7). The ratio between the two bands, and their migration distance relative to that of bromophenol blue, when electrophoresed in the presence of high, unphysiological concentrations of calcium suggests that the bands observed by Rusca and Calissano (11) are not related to the two components described by us (Fig. 1). This

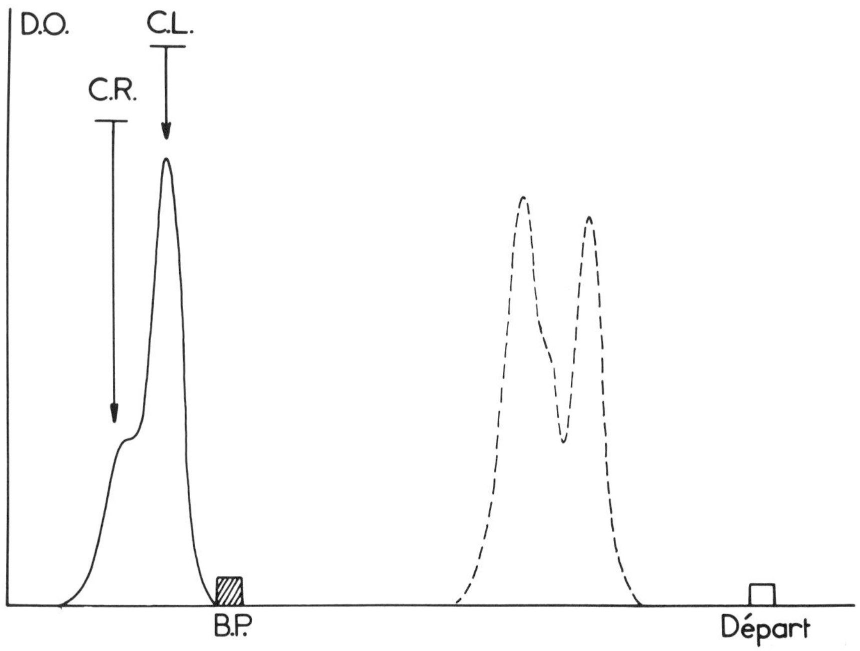

Fig. 1. Densitometric profiles of pig F-S100 after electrophoresis on mixed (7 %) acrylamide- (0.75 %) agarose gel slabs :
——— 90 mM Tris-acetate (pH 8.6), 2.5 mM EDTA ; ----- 90 mM Tris-acetate (pH 8.6), 4 mM $CaCl_2$.

conclusion is also supported by the fact that the protein fractions induced by high calcium concentration were partially separated by gel filtration (11), where ours were not separated (7). Electrophoregrams of purified pig brain S-100 in the presence of lower Ca^{++} concentrations (between 10 and 400 µM) (Fig. 2) were qualitatively similar to these on gels containing neither Ca^{++} nor EDTA (7, 8). In fact, pig brain S-100 resolved as up to five bands, as did beef brain S-100. The migration of the whole S-100 electrophoretic pattern relative to that of bromophenol blue was progressively reduced as the calcium concentration increases (7).

A puzzling result which casts serious doubts on the hypothesis that increased brain calcium is responsible for the electrophoretic heterogeneity of rat brain F-S100 protein was given by electrophoresis of brain soluble proteins of the three species examined in

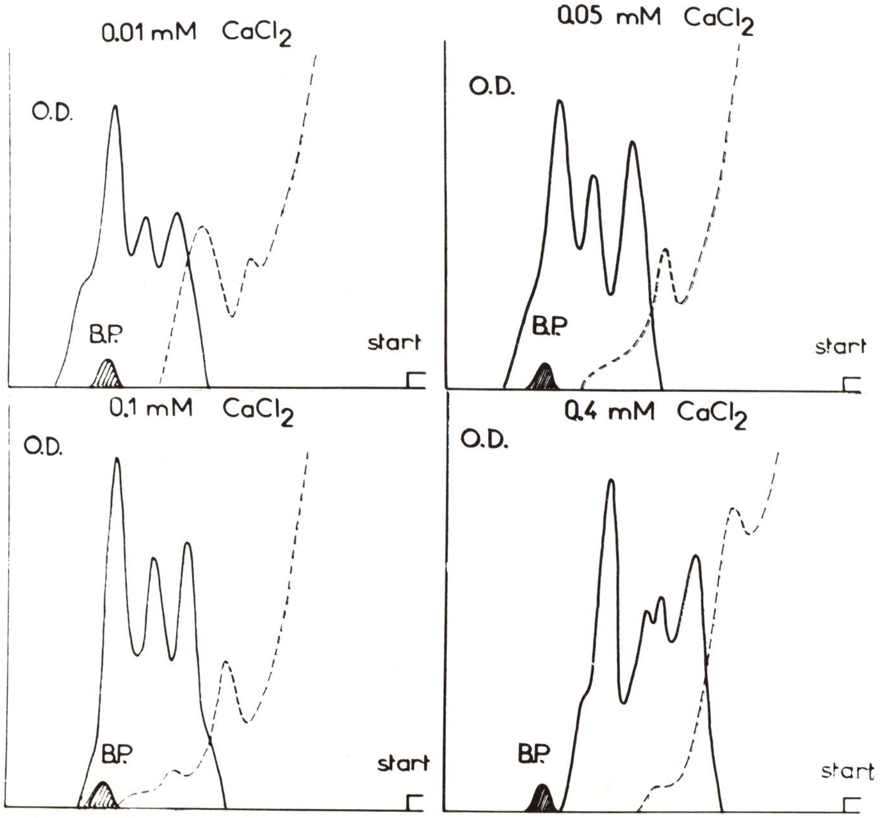

Fig.2. Densitometric profiles after electrophoresis on mixed (7 %) acrylamide- (0.75 %) agarose gel slabs in 90 mM Tris-acetate (pH 8.6) containing different concentrations of $CaCl_2$. ——— pig F-S100; ------ total pig brain soluble proteins ; BP : bromophenol blue.

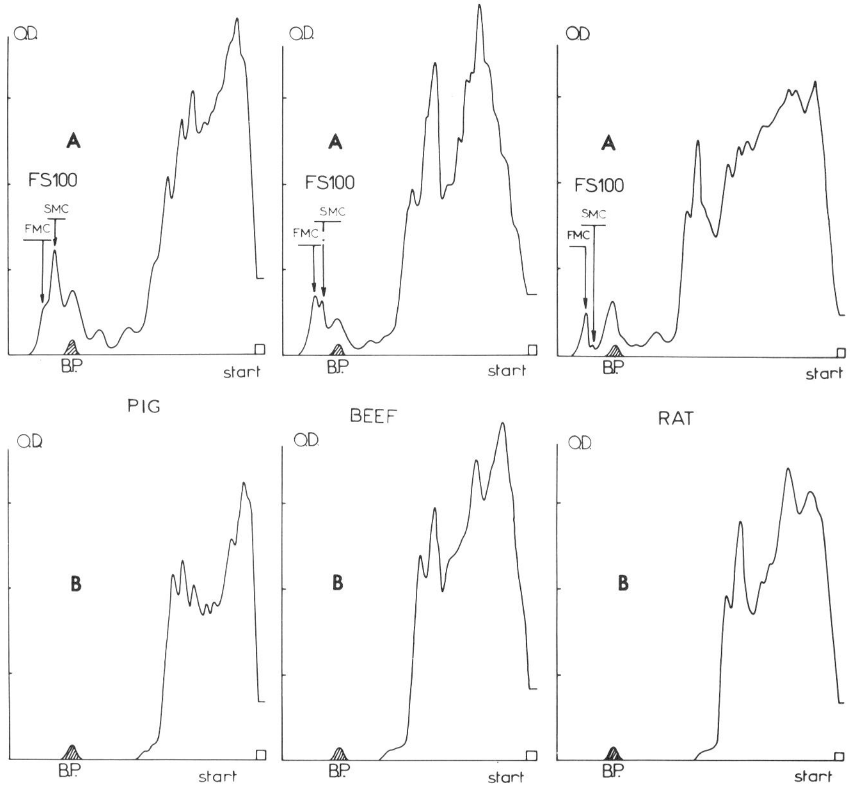

Fig. 3. Densitometric profiles after electrophoresis on mixed (7 %) acrylamide- (0.75 %) agarose gel slabs of total brain soluble proteins in A : 90 mM Tris-acetate (pH 8.6), 2.5 mM EDTA ; B : 90 mM Tris-acetate (pH 8.6), 4 mM $CaCl_2$.

the presence of high (Fig. 3) and low (Fig. 2) Ca^{++} concentrations. None of the normal prealbumin bands, (including S-100) were visible (7). A comparison of the migration distance of purified S-100 relative to that of bromophenol blue showed that under these conditions the protein migrated to the same position as many other brain soluble proteins and for this reason, it could not be detected on electrophoresis of brain extracts (Fig. 3) (7). On the basis of these results it is reasonable to conclude that FMC and SMC are not related to calcium action. Thus they must be either two different conformational states of a single protein (not due to calcium action) or two (at least) different molecular species. The immunological similarity of the two components is not an obstacle to the latter hypothesis ; in fact differences between F-S100 of different species shown by amino acid analysis and electrophoresis were not detected immunologically.

Only a structural analysis of the two components can give a direct proof of the molecular heterogeneity of F-S100. A partial separation of the two components of pig brain F-S100 was obtained (Fig. 4) by using chromatography on DEAE-cellulose, with a very shallow NaCl gradient (8).

F-S100 was eluted as a non-symmetrical peak. The "shoulders" on the elution profile were due to the partial separation of the two components.

After electrophoretic control the various tubes were pooled in three fractions :

- fraction I contained only slow migrating component ;
- fraction II contained almost exclusively slow migrating component ;
- fraction III was a mixture of the two components. In addition control electrophoresis of this fraction revealed the presence of a number of bands, migrating behind F-S100, We believe that these

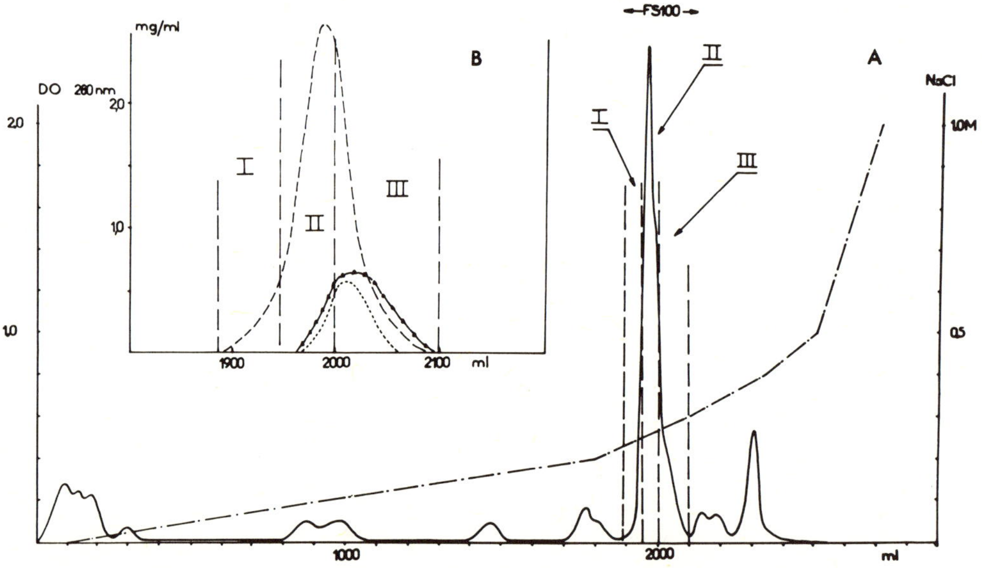

Fig. 4. DEAE-cellulose column chromatography of an F-S100 rich fraction obtained by ammonium sulphate fractionation (8). A:elution profile ——— protein ; B : distribution of FMC (.....), SMC (-----); and partially denatured F-S100 (-Δ-Δ-Δ-).

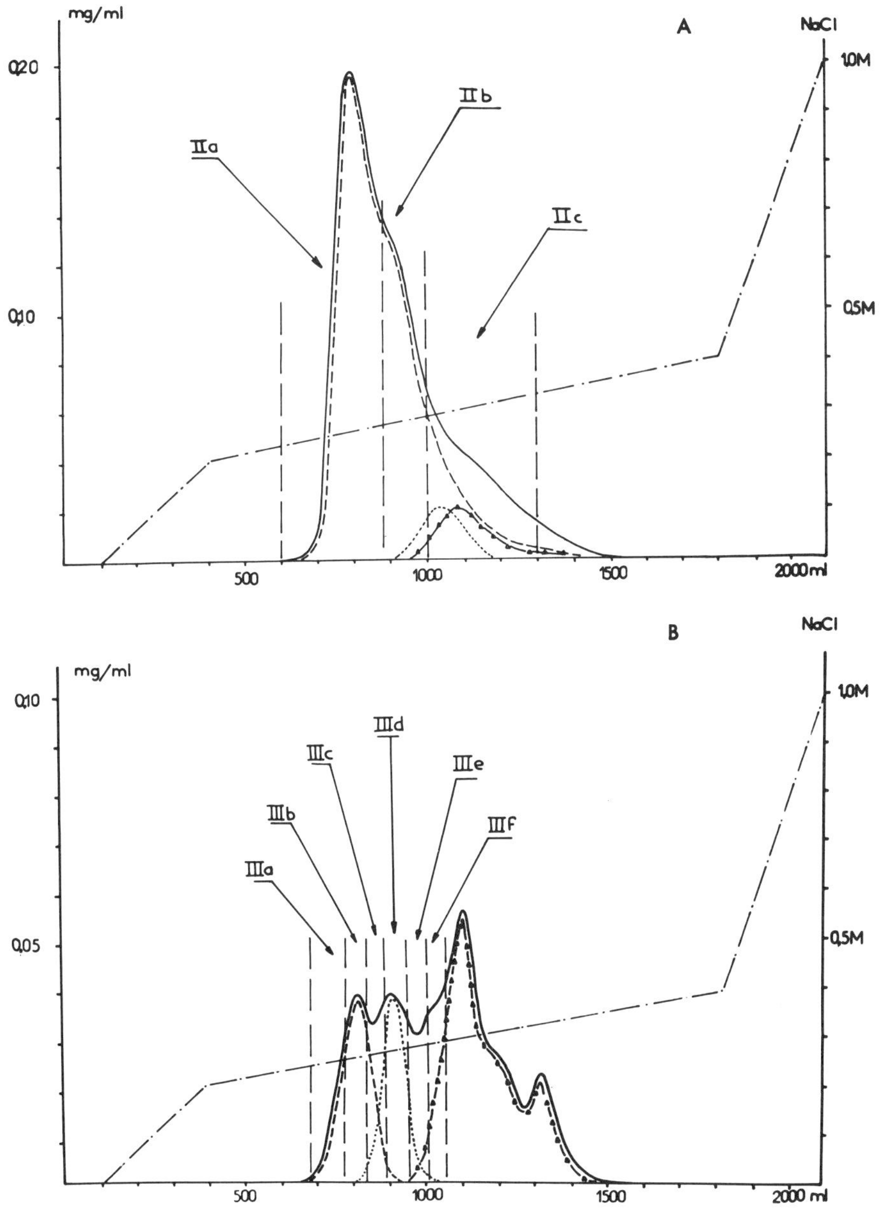

Fig. 5. Rechromatography on DEAE-cellulose of fraction II (A) and fraction III (B) of F-S100 as defined in Fig. 4. ——— protein ; ------ SMC ; ...... FMC ; —Δ—Δ—Δ— partially denatured F-S100.

bands represent denaturated F-S100 since their migration distance relative to that of bromophenol-blue, was similar to those of F-S100 electrophoresed without EDTA (8). Furthermore, samples of F-S100 allowed to stand at room temperature for a few days, or precipitated by HCl and then dissolved in Tris-acetate-EDTA (pH 8.6), showed, upon electrophoresis, similar bands although such bands were not found in the original samples (14).

Fractions II and III were rechromatographed on DEAE-cellulose with very shallow gradients (8). As can be seen in figure 5, we obtained from fraction 2 one subfraction (fraction IIa) containing only slow migrating component. Fraction III gave several subfractions, one of which contained practically only fast migrating component (fraction IIId).

The tryptic peptide maps of the pooled SMC fractions compared to that of the FMC fraction are shown in Fig. 6 (8). A number of

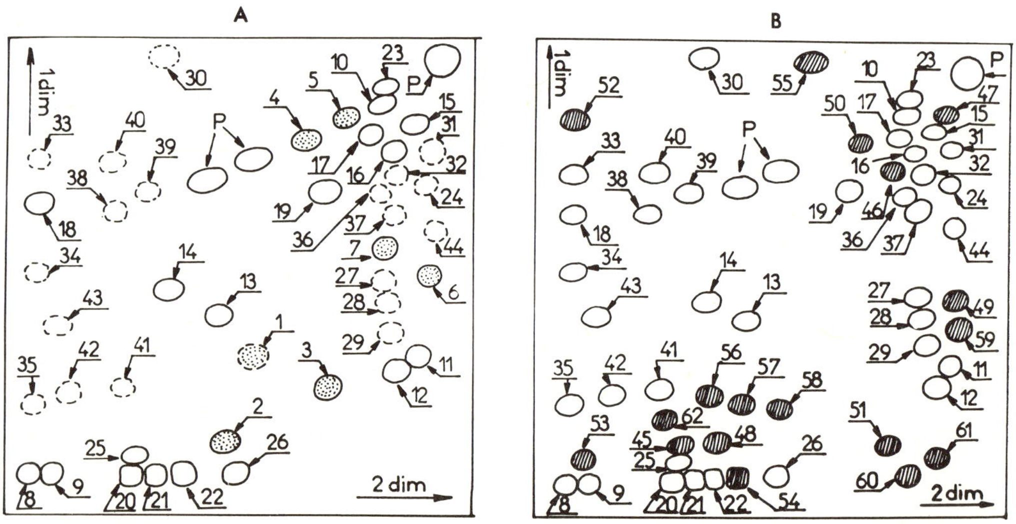

Fig. 6. Silica gel thin-layer chromatograms of dansylated tryptic hydrolysates of pig F-S100 FMC and SMC. A: FMC ; B: SMC ; peptides specific to FMC ; peptides specific to SMC ; common peptides ; P : breakdown products of dansyl chloride weakly fluorscent spots are defined by a broken line. The material used was derived from 100 µg protein. 1st dimension : methyl acetate-isopropanol-25 % ammonia (9:6:4, by vol) ; 2nd dimension : isobutanol-acetic acid-water (15:4:2, by vol).

peptides spots were common to the maps of both fractions. Many others were specific to one of the two components, suggesting that either FMC and SMC are indeed two different molecules having some similarities ("isoproteins") or that F-S100 contains proteins, other than the S-100 antigen, but with similar chromatographic and electrophoretic characteristics. Contamination with proteins migrating differently cannot be more than 2 % taking into account the amount of S-100 used for control electrophoresis and the sensitivity of Coomassie brilliant blue staining.

It appears however that SMC contains by itself more than one molecular species, since the number of spots in the maps was very different and greater for the SMC map, and since subfractions of SMC have different fingerprints (Fig. 7) (8).

It is evident that these conclusions depend on the absence of peptides due to non-specific cleavage, and on the absence of chromatographic artifacts such as double spots for a single peptide. Both sources of errors can be ruled out. In fact the specificity of tryptic digestion was checked by finger-printing known proteins (16) and by analysis of the C-terminal amino acid of each peptide, which indicated that this amino acid was lysine or arginine for almost all the peptides. Chromatographic artifacts were ruled out by eluting the spots after the first dimension and evaporating the samples. Each spot was then chromatographed in the second dimension. The reconstituted map of whole F-S100 showed the same high number of peptide spots.

This number is higher than that expected from the lysine and arginine residues present in the whole F-S100 protein fraction determined on the same preparation of F-S100. Since we have excluded contamination with other proteins and technical artifacts, the high number of peptides is compatible only with the presence of several molecular species. In view of the fact that S-100 appears to contain at least 3 polypeptide chains (3), we suggest that F-S100 isoproteins are due to various combinations, in sets of three, of the constituent polypeptide chains. There would be more than three chains differing from each other, in some cases by "fine" punctual, and in other cases by more gross differences of their primary structure. This suggested "molecular heterogeneity of S-100" is of the same type of that of normal haemoglobin. Regional (4, 5) and interspecific differences (12) of F-S100 could also be explained within the same framework. The various isoproteins would have similar external surfaces in order to explain the immunological results. In the presence of EDTA, the various isoproteins could be separated into two classes, that we called FMC and SMC. It is possible that Ca^{++} unfolds these molecules, and as suggested by Calissano, Moore and Friesen (1969) exposes hidden amino acid differences between the various molecules, thus leading to the multiple band electrophoretic

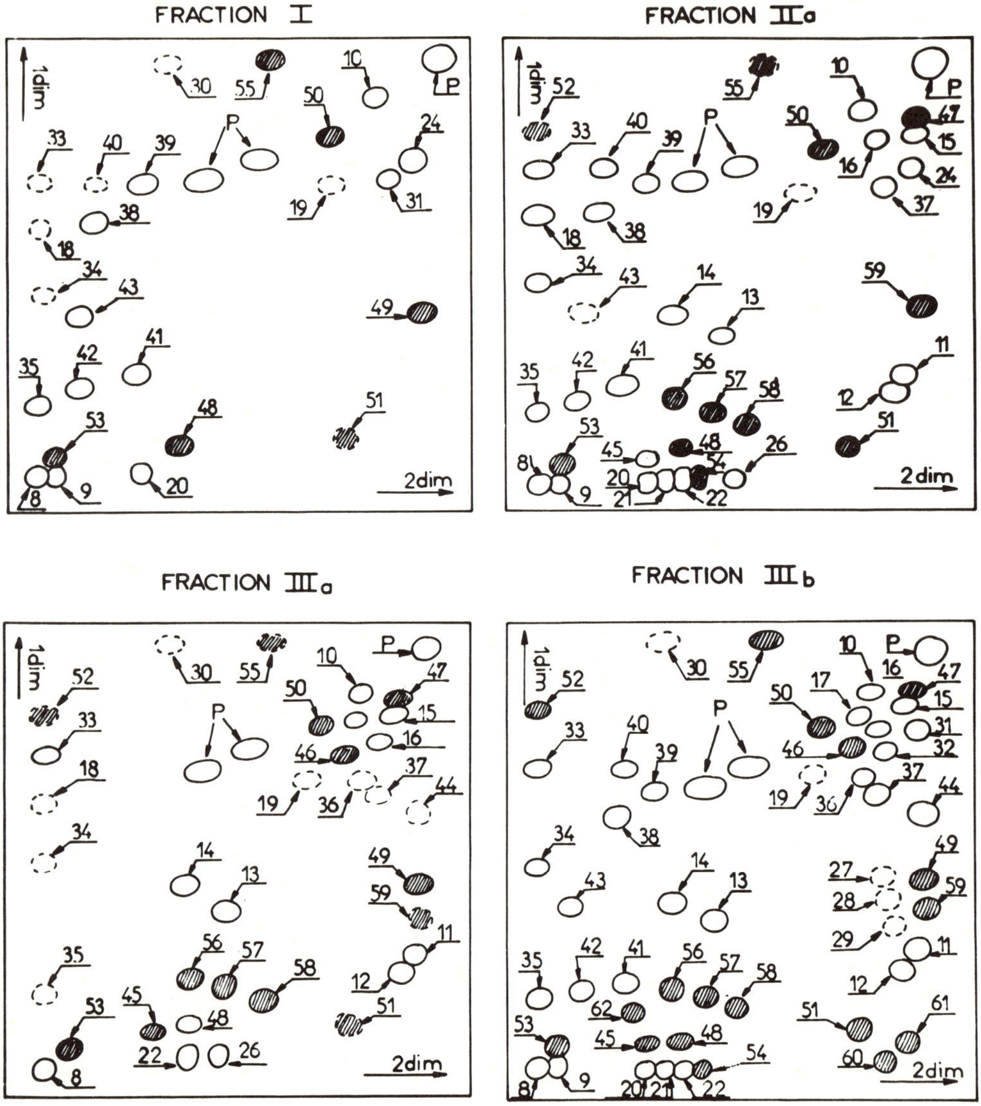

Fig. 7. Silica gel thin-layer chromatography of dansylated tryptic hydrolysates of four fractions (as defined in Fig. 4 and 5) which contained only SMC. For conventions and conditions see legend to Fig. 6.

profiles. High Ca^{++} concentration could expose hydrophobic residues (Ile, Leu and Val account for at least 20 percent of amino acids in pig brain F-S100) and thus reduce F-S100 solubility and its electrophoretic mobility.

It is impossible at the present time to relate the extensive molecular heterogeneity of F-S100 to function, since this function, despite the many hypotheses advanced, is unknown. Studies attempting to elucidate the functional role of this protein, however, should take into account the existence of these isoproteins.

ACKNOWLEDGEMENTS

G.Gombos : Attaché de Recherche au CNRS. We thank A.Meyer and R. Langs for their technical assistance. This work was in part supported by the Institut National de la Santé et de la Recherche Médicale (contract 7111698) and the Fondation pour la Recherche Médicale Française.

REFERENCES

1 CALISSANO, P., MOORE, B.W. and FRIESEN, A., Biochemistry, 8 (1969) 4318.
2 CALISSANO, P. and RUSCA, G. in 2nd Intern. Meeting Intern. Soc. Neurochem. (Ed. PAOLETTI, R., FUMAGALLI, R. and GALLI, R.) Publ. Tamburini, Milan, 1969, p.115.
3 DANNIES, P.S. and LEVINE, L., Biochem. Biophys. Res. Commun., 37 (1969) 587.
4 FILIPOWICZ, W., VINCENDON, G., MANDEL, P. and GOMBOS, G., Life Sciences, 7 (1968) 1243.
5 GOMBOS, G., FILIPOWICZ, W. and VINCENDON, G., Brain Res., 26 (1971) 475.
6 GOMBOS, G., VINCENDON, G., TARDY, J. and MANDEL, P., C.R.Acad. Sc. (Paris) Sér.D, 263 (1966) 1533.
7 GOMBOS, G., ZANETTA, J.P., MANDEL, P. and VINCENDON, G., Biochimie, 53 (1971) 635.
8 GOMBOS, G., ZANETTA, J.P., MANDEL, P. and VINCENDON, G., Biochimie, 53 (1971) 645.
9 LEVINE, L. and MOORE, B.W., Neurosci. Res. Progr. Bull., 3 (1965) 18.
10 MOORE, B.W., Biochem. Biophys. Res. Commun., 19 (1965) 739.
11 RUSCA, G. and CALISSANO, P., Biochim. Biophys. Acta, 221 (1970) 74.
12 TARDY, J., GOMBOS,G., VINCENDON, G. and MANDEL, P., C.R.Acad. Sc. (Paris) Sér. D, 267 (1968) 669.
13 UYEMURA, K., TARDY, J., VINCENDON, G., MANDEL, P. and GOMBOS,G., C.R. Soc. Biol. Fr., 161 (1967) 1396.
14 UYEMURA, K., VINCENDON, G., GOMBOS, G. and MANDEL, P., J. Neurochem., 18 (1971) 429.
15 VINCENDON, G., WAKSMAN, A., UYEMURA, K., TARDY, J. and GOMBOS,G., Arch. Biochem. Biophys., 120 (1967) 233.
16 ZANETTA, J.P., VINCENDON, G., MANDEL, P. and GOMBOS, G., J.Chromatog., 51 (1970) 441.

IMMUNOLOGICAL STUDIES OF BRAIN SPECIFIC PROTEIN

Krystyna WARECKA

Nervenklinik der Medizinischen Akademie Lübeck (24)
(Germany)

Detection of the Protein

We have found a glycoprotein in human brain which is brain specific on immunological criteria (7). As can be seen in Fig.1A, immune sera prepared against white matter, give precipitation bands with aqueous extracts from white matter. Only one band remains if this serum is absorbed with human serum and organ extracts (Fig.1B).

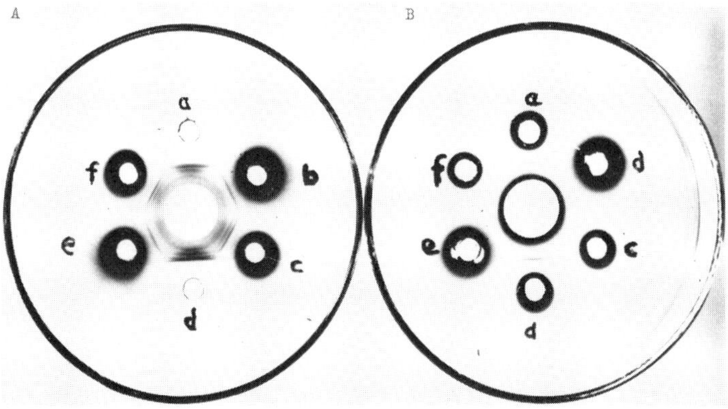

Fig. 1. A. Immunoprecipitation of anti-white matter immune serum against white matter with aqueous extracts of white matter (a and d), liver (b and e), and kidney (c and f). Two distinct precipitation lines are seen, one of which shows crossing-over in the bordline area between white matter-liver and white matter-kidney.
B. Immunoprecipitation after absorption of anti-brain serum with aqueous liver and kidney extracts. Only one brain-specific line remains.

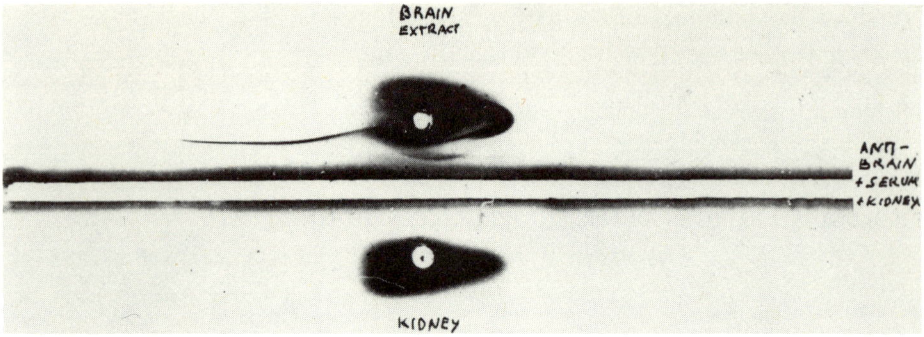

Fig. 2. Immunoelectrophoresis of brain and kidney extracts (pH 6.5-8.2) in the $alpha_2$-area. It appears as a sharply defined, rather straight precipitation line. At its cathodic end, the line terminates in a semi-circle around the application point. Upper hole : white matter extract ; lower hole : kidney ; center well : antibrain serum absorbed with human serum and liver.

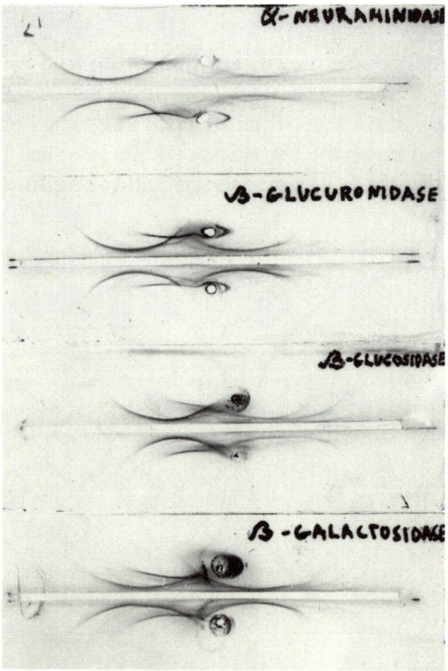

Fig. 3. Immunoelectrophoresis of enzyme-treated antigens. When neuramidase was added to the antigen, this line became shorter and less intense, the semi-circle around the application point disappeared. Control tests with glucuronidase, glucosidase and galactosidase show less definite alterations. Upper hole in a,b,c,d : brain extract incubated with enzymes ; lower hole : control brain extract ; center well : antibrain serum.

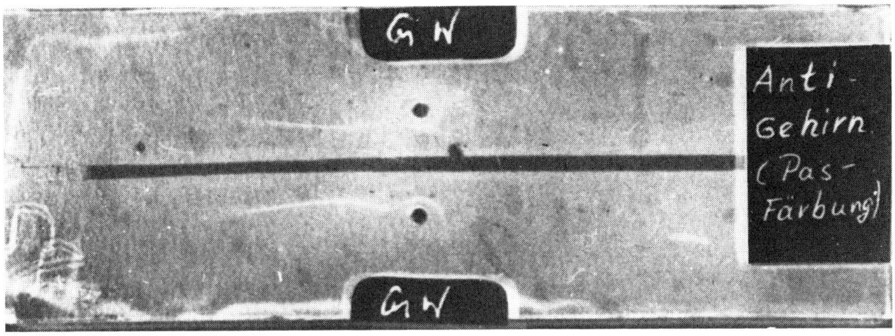

Fig. 4. Immunoelectrophoresis of aqueous extracts from white matter (GW) and antibrain serum. PAS stain. The $alpha_2$-precipitation line is the only fraction which is clearly visible.

After absorption of the immune serum with liver and kidney extracts, only the brain specific band can be seen. The protein migrates in the $alpha_2$ area, and by immunoelectrophoretic analysis gives a sharply defined, rather straight precipitation line which terminates as a semi-circle around the application well (Fig. 2).

The nature of this protein has been examined by enzymatic digestion (Fig. 3) and periodic acid-Schiff staining (Fig. 4). The immunoelectrophoretic analysis of the protein was affected by prior treatment with α-neuraminase, whereas treatment with β-glucuronidase, β-glucosidase or β-galactosidase had little or no effect. When the immunoelectrophoregram was stained by the periodic acid-Schiff method only the $alpha_2$-antigen precipitation line was visible.

From the totality of these results, several deductions as to the nature of the protein can be made. At neutral pH, the protein is acidic because even in acidic buffers it migrates towards the anode. A part of this negative charge may be due to the presence of neuraminic acid. The acidic charge of the protein is confirmed by its behaviour on ion exchange chromatography (see below). The protein is clearly a glycoprotein, since it contains neuraminic acid and stains with the periodic acid-Schiff reaction.

Cellular and Subcellular Localization

The subcellular localization of the protein has been studied using the fractionation procedure of Gray and Whittaker (2). All fractions were examined by electron microscopy to check their purity, then subjected to immunological analysis. The microsomal fraction seemed to be highly pure (Fig. 5). When the fractions were subjected to immunological analysis, precipitation lines were obtained with the microsomal and soluble fractions but no reaction was ob-

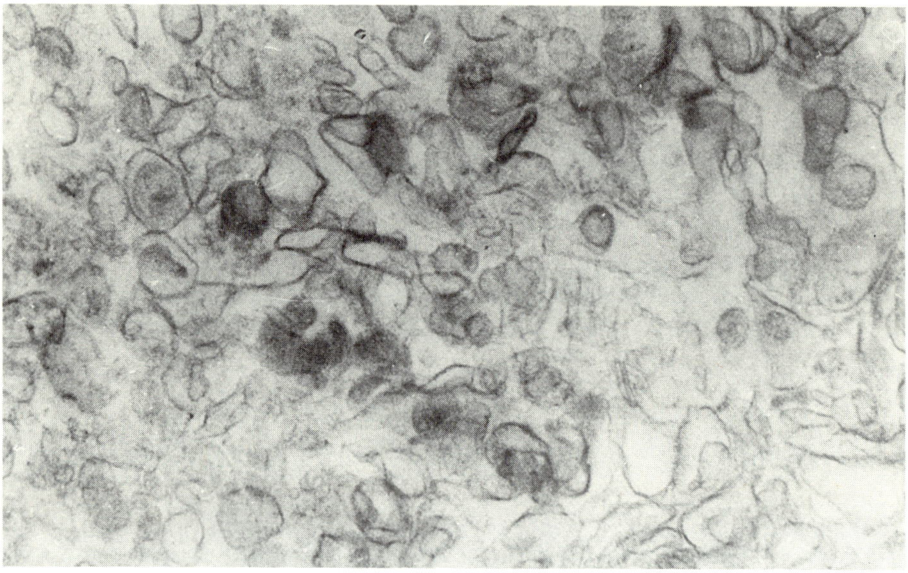

Fig. 5. Microsomal fraction. Vesicles of various sizes and small filaments corresponding to endoplasmatic reticulum are seen. 66,400 X.

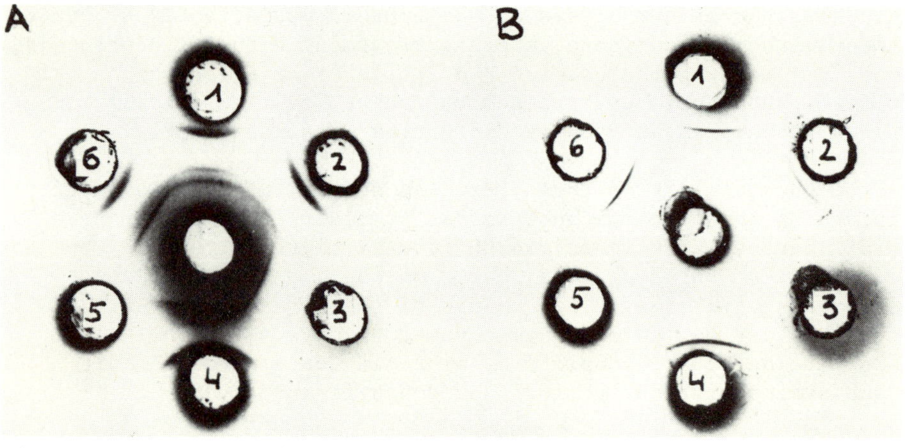

Fig. 6. Immunodiffusion analysis : 1 and 4 brain extract ; 2 cell sap ; 3 liver ; 5 human serum ; 6 microsomes ; center well : anti-brain serum before (A) and after (B) absorption with liver and human serum.

tained with the nuclear, myelin, synaptosomal or mitochondrial fractions (Fig. 6).

Comparative studies on rat, monkey and human brain indicate a "phylogenetic development" of this protein : the human brain protein shows semi-identity with a rat brain specific protein and complete, or almost complete,identity with a monkey (Macaca mulatta) brain specific protein (5).

This human brain specific protein appears at a specific stage in brain development (Table I). The protein can be first detected between 24 and 28 weeks of foetal age, when glial cells differentiate ; at the "myelination glia stage" (Fig. 7). It is at this time that myelin, astrocytes, macro- and micro-glia, and the first oligodendrocytes can be identified histologically. Detection of the brain specific protein by immunoelectrophoresis was less specific than detection by the induction of antibody synthesis since immunologically the protein appeared 3-4 months after birth whereas by antibody induction, it could be detected at 24-28 weeks of foetal age.

The fact that the brain-specific protein appears during neurogenesis in parallel with the myelination glia suggested that the brain specific alpha$_2$-glycoprotein is localized in glia cells. This hypothesis is supported by the fact that white matter extracts are more effective in inducing the specific antibody formation than are gray matter extracts. Furthermore this protein is present in higher concentrations in normal and pathological white matter than in gray matter. Only trace amounts of this protein are found in myelin-poor regions such as the putamen and caudate nucleus. The glial localization of this protein was tested directly on neurons and glial cells isolated by the density gradient method of Rose (3), slightly modified (12). Four fractions (debris, neuropil, nuclei and neuronal perikarya) were obtained. The nuclei, neuropil and neuronal perikarya were examined immunologically, and by light and phase contrast microscopy.

The morphology of the fractions obtained is shown in Fig. 8. Some contamination of the neuronal perikarya fraction with glial cells was detected. A quantitative estimate of the contamination was obtained by counting the number of glial cells, neuronal perikarya and neuronal nuclei without surrounding cytoplasm. The intactness of the cells was defined by the neuronal perikarya/neuronal nuclei ratio (Table II). The contamination of the neuronal perikary fraction with glial cells and the neuropil fraction with neuronal elements was estimated from the ratio of neuronal perikarya + neuronal nuclei/glial cells. The results show that both the neuronal perikarya and neuropil fractions are highly enriched (Table II). In the neuropil fraction predominantly glial cells are seen.

When the fractions were examined immunologically (Fig. 9.),the brain specific precipitation band was clearly visible with extracts of the neuropil fraction. The precipitation band had the same immuno-

TABLE I

Results of Immunological and Histological Studies in 25 Fetal and 10 Infantile Brains

Length of Fetus cm	Weight of Fetus g	Weight of Brain g	Age			Immunoelectrophoresis of extracts of Fetus brain tested with anti-brain-serum			Antibodies against brain spec. protein	Histological examination	Number of cases examined
			Weeks	Months	Years	Serum prot.	Organ prot.	Brain spec. prot.			
0.8	1.5		3-4			+	-	-	no material	no material	1
7.0	21.8	3.80	8-9			+	-	-	-		1
9.0	23.6	4.02	10-11			+	-	-	-	Neuroblast Glioblast	1
12.0	35.0	7.70	12-13			+	+	-	not tested		1
16.0	103.2	15.45	16-18			+	+	-	-	Neuroblast, Glioblast (nerve cells) (macroglia)	2
37.0	1082.0	174.00	24-28			+	+	-	+	Myelination glia in the cerebrum, thin myelin sheaths in the pons	9†
46.0	1700.0	250.00	32-36			+	+	-	+	Myelination glia in the cerebrum, thin myelin sheaths in the midbrain	8††

54.0	3445.0	330.00	40-42 birth			+	+	-	+	Myelination glia, myelin and differentiation of glia	2
		750.00		3-4		+	+	±	+	Myelination glia, myelin and differentiation of glia	2
		1200.00			1-5	+	+	+	+	Brain fully developed	8

†Five of these survived up to three days ; ††Five of these survived up to 17 days. In those collectives, in which more than one brain was examined, the length and weights are given as an average.

In the immunoelectrophoresis a small precipitation line indicating the appearance of this protein was first seen in the brains of 2 newborn children, who survived 3-4 months. After this period, the alpha$_2$-glycoprotein was found consistently in 8 further specimens of any age group ranging from 1 to 5 years. On the other hand rabbits injected with foetal brains of the foetal period of about 24-28 weeks and above 28 weeks produced antibody to alpha$_2$-glycoprotein. The same results were obtained by immunization with brains of children of the age of 3-4 months of life and 1-5 years. However rabbits immunized with material from foetal brains of age of 8-9 weeks, 10-11 weeks and 16-18 weeks did not produce antibodies to brain-specific glycoprotein.

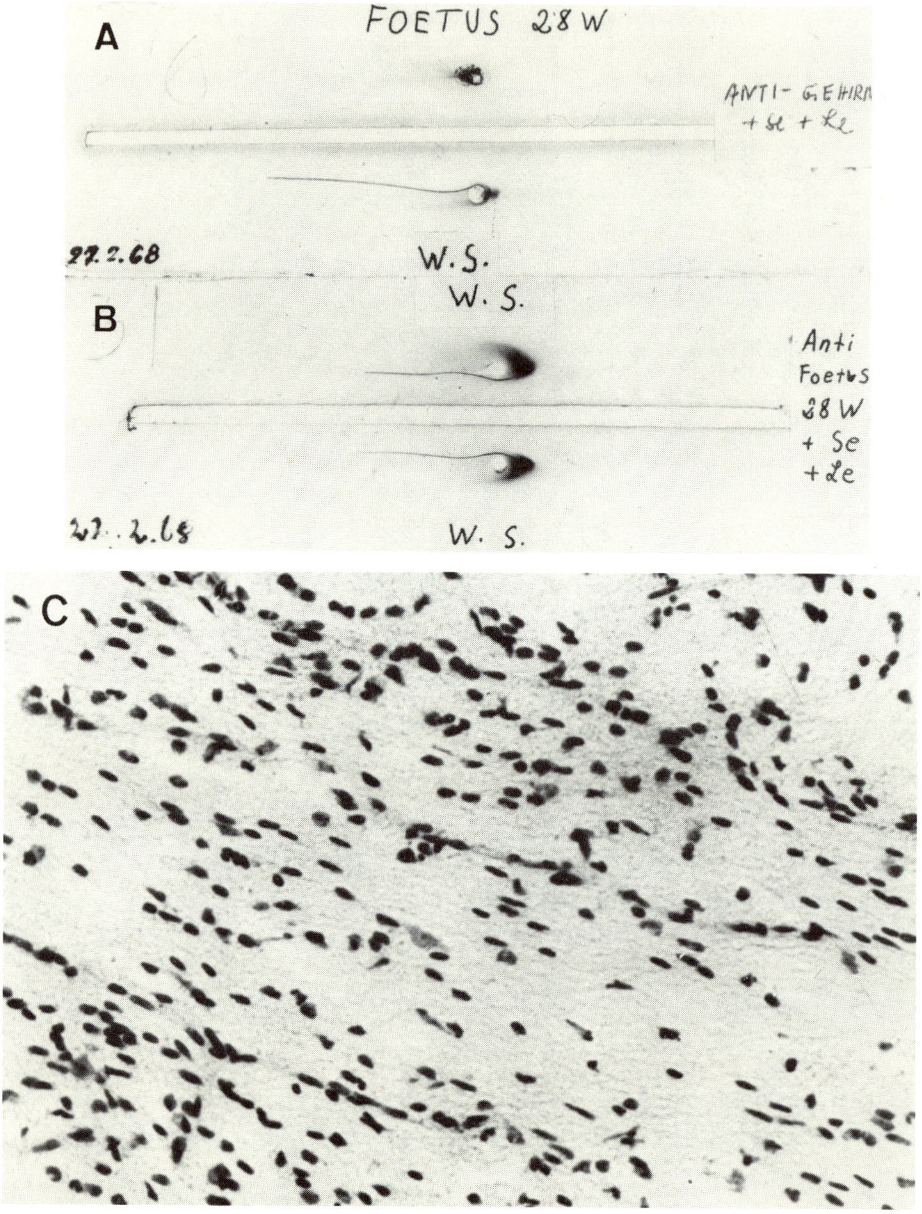

Fig. 7. A. Immunoelectrophoresis of extract of brain of a 28 week-old foetus (upper hole) and extract of normal adult brain (lower hole) developed with antibrain serum absorbed with human serum and human liver. - B. Immunoelectrophoresis of normal adult brain extract (upper and lower hole) developed with antiserum against 28 week-old foetal brain, absorbed with human serum and human liver. C. Myelination glia in the internal capsule between the globus pallidus and corpus luysii of the same foetus. Cresyl violet ; 215 X.

TABLE II

Fraction	Neuronal perikarya / neuronal nuclei	Neuronal perikarya+neuronal nuclei / glial cells
	WHITE MATTER	
Suspension	0.32	0.64
Neuropil	0.03	0.08
Neuronal	2.05	3.15
Nuclear	0.24	1.705
	GRAY MATTER	
Suspension	0.48	0.62
Neuropil	0.005	0.055
Neuronal	2.11	3.33
Nuclear	0.26	0.62

The measure of intactness of the isolated cell preparations was defined on the basis of the neuronal/nuclear ratio. The contamination of glial fraction with neurons and neuronal fraction with glial cells in the white and gray matter has been calculated by neuronal/glial ratio. In this ratio the term neuronal includes the intact neuron as well as the nucleus. Suspension means initial tissue suspension in Ficoll medium (10 % Ficoll, 100 mM potassium chloride and 10 mM potassium buffer pH 7.4).

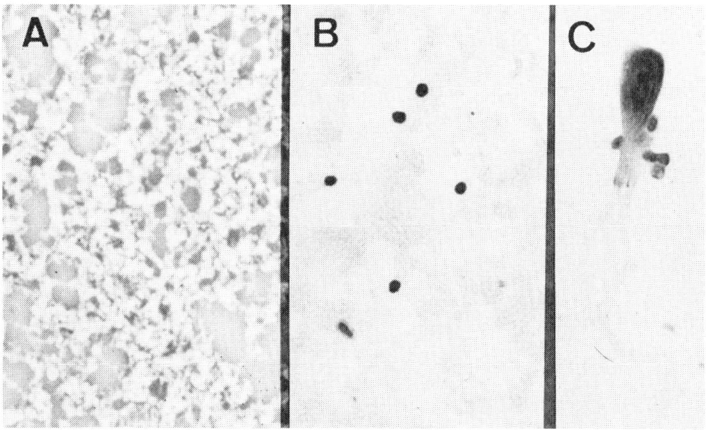

Fig. 8. Samples were examined either by phase contrast or after light staining with toluidine blue by light microscopy. A and B : glial cells in phase contrast and light microscopy respectively ; C : glial cells attached to neuron, small amounts of glial contamination of the neuronal fraction was noticeable. 400 X.

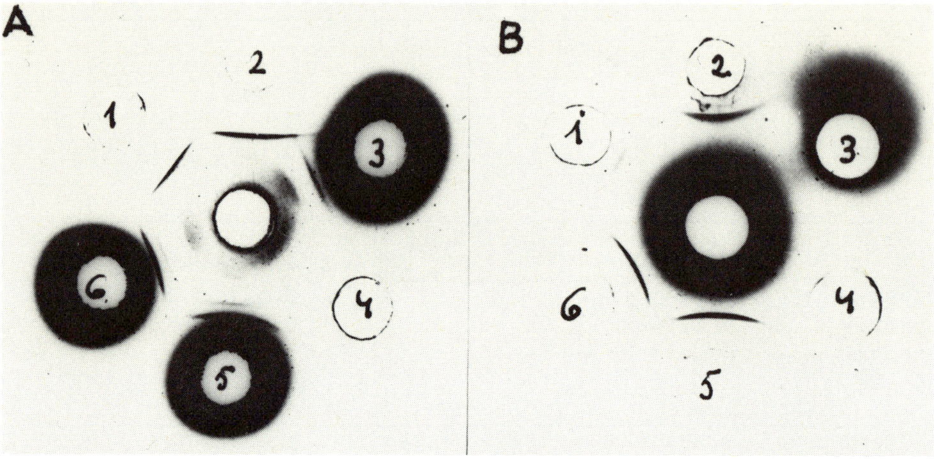

Fig. 9. Immunodiffusion analysis of glial fraction (1 and 2), liver (3), serum (4), and brain (5 and 6) with antibrain serum (center well) before (A) and after (B) absorption with liver and human serum. The brain-specific precipitation line is distinctly visible.

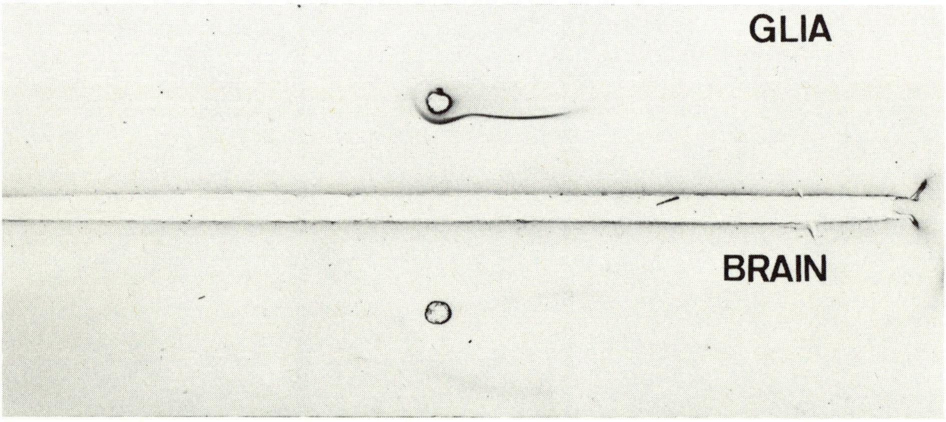

Fig. 10. Immunoelectrophoresis of glia (upper hole) and brain tissue (lower hole) developed with antibrain serum absorbed with liver and human serum (center well).

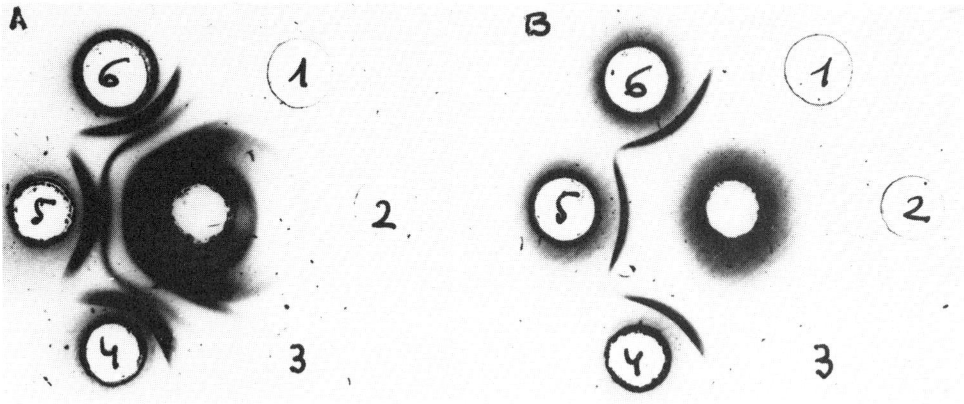

Fig. 11. Immunodiffusion analysis of neuronal fraction (1, 2, 3) and brain tissue (4, 5, 6) with antibrain serum (center well) before (A) and after (B) absorption with liver and human serum. No precipitation lines related to brain-specific protein are seen in the neuronal fraction. The neuronal protein in every well amounted to 242 μg.

electrophoretic properties as the alpha$_2$-protein detected in white matter (Fig. 10). Analogous experiments with neuronal extracts (Fig. 11) and the nuclei did not show precipitation bands although the amount of protein applied from these two fractions was 10-fold higher than that necessary to detect the protein with the neuropil fractions.

Isolation of the Alpha$_2$-Glycoprotein

A method has now been developed in our laboratory for the purification of the brain-specific alpha$_2$-glycoprotein. The protein was retained by DEAE-cellulose, and could be eluted by acidic buffers of high ionic strength. A fraction highly enriched in the protein (Fraction IV) could be obtained by elution with a discontinuous gradient of phosphate buffer-NaCl (Fig. 12). Fraction IV showed only one immuno-precipitation band with antibrain immune serum (Fig.13), but if the isolated protein was used to induce antibody, there was some trace reaction with serum albumin.

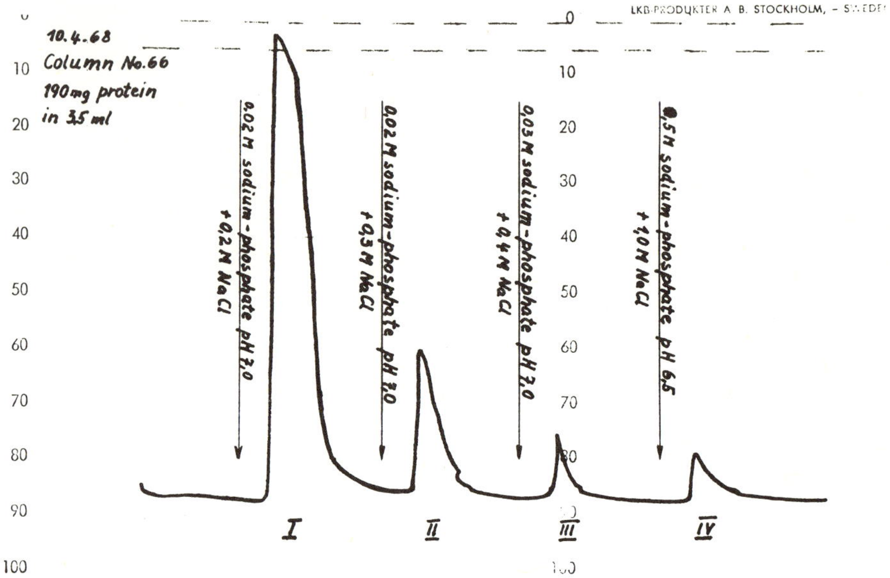

Fig. 12. The alpha$_2$-brain-specific protein was first isolated on DEAE-cellulose by using a phosphate buffer with increasing activity and molarity (6). The brain-specific protein was eluted in the fraction IV.

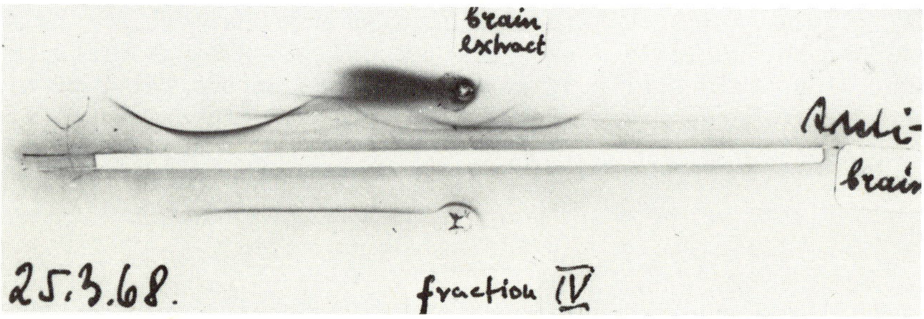

Fig. 13. Immunoelectrophoresis of brain extract (upper hole) and fraction IV (lower hole) with antibrain serum (center well).

To continue biological studies of the protein, homologous antibodies were needed. Such antibodies were isolated by affinity chromatography (4) as described by Axen (12). Antibrain immune serum was passed through a Sepharose-liver column and the unretarded material collected. This eluate was then passed through a Sepharose-brain column, and after elution of the unretarded material, the

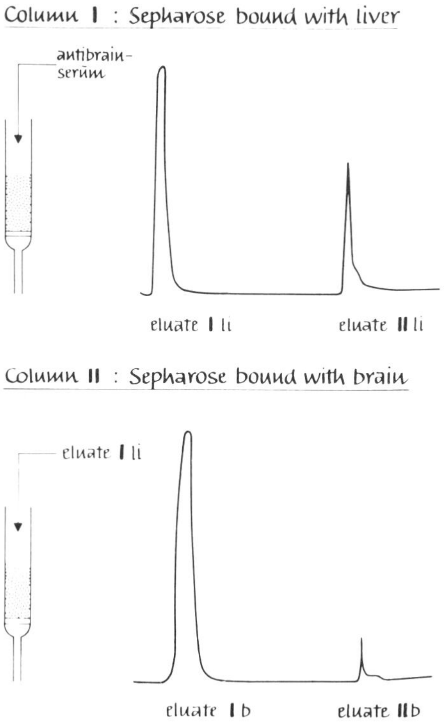

Fig. 14. Antibrain serum obtained by immunization of rabbits with brain extract was added to Sepharose-liver column. A sample (100 ml) of 0.15 M phosphate buffer pH 7.2 was then passed through the column at a rate of 45 ml/h. The antibodies bound to the immunoadsorbent were eluted with 0.1 M acetic acid pH 3.0 (fraction II li). The first eluate of the Sepharose-liver column was added to Sepharose-brain column. After 100 ml of buffer had passed, acetic acid was added to elute the bound antibodies (fraction II b). Details see Ref. (6).

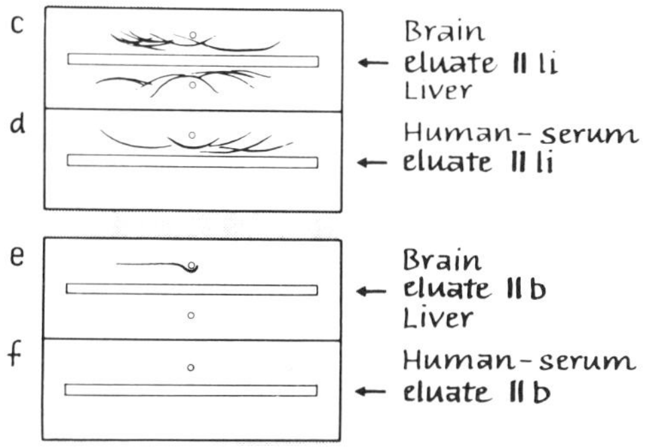

Fig. 15. The fraction II li contains no brain-specific antibodies, no brain-specific line is here visible (c and d). The fraction IIb is completely pure, reacts only with brain-specific protein (e and f).

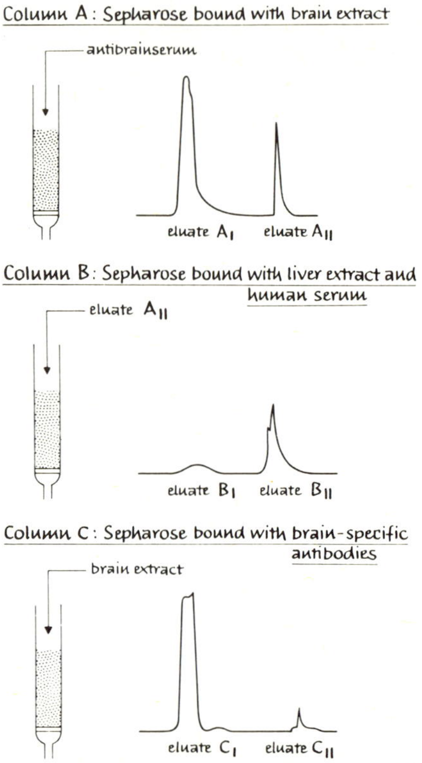

Fig. 16. Antibrain serum (10 ml containing 0.7 g protein) was passed through the column (column A = 9 x 16 cm). After the column was washed with 0.15 M phosphate buffer pH 7.2 the bound antibodies were eluted by 0.15 M glycine-HCl buffer pH 1.8 (eluate A_{II}). This eluate was adjusted at pH 7.2 with 1 N NaOH and passed through the Sepharose liver column (column B = 14 x 1.6 cm). The eluate B_I was bound to Sepharose (column C = 4 x 2 cm). Brain extract (10 ml containing 50 mg protein) was passed through this column. The column was washed and then eluted with glycine-HCl buffer. The eluate C_{II} contains only brain-specific proteins. The yield of the brain-specific proteins was 0.13 mg.

bound antibodies were eluted with 0.1 M acetic acid (pH 3.0), as shown in Fig. 14. This eluate reacted only with brain, giving the typical immuno-electrophoretic pattern of alpha$_2$-glycoprotein. No reaction was seen with liver or serum (Fig. 15).

This principle was also used for isolating brain-specific proteins. Antibrain serum was passed through a Sepharose-brain extract column and the column washed (Fig. 16). The bound antibodies were eluted with 0.15 M glycine-HCl (pH 1.8). The eluted material was then applied to a Sepharose-liver extract and human serum column.

The column was then washed and the material washed from the column bound to Sepharose to form a Sepharose-brain specific antibody column. Brain extract was then passed through the column, the column was exhaustively washed and the bound brain-specific proteins eluted with 0.15 M glycine-HCl (pH 1.8). This protein fraction was analysed immunologically as shown in Fig. 17. Eluate C_{II} shows one, and sometimes two, precipitation bands which are identical with the brain specific bands. By immunoelectrophoresis the bands were also visible, occasionally the second minor band was visible (Fig. 18). As shown in Fig. 19 and 20, anti-C_{II} immune serum reacts with brain extract and fraction C_{II} but neither with liver nor serum.

At present, we can only speculate as to the nature of the second brain-specific protein. But it seems likely that this protein belongs to the same family of brain specific proteins as does the alpha$_2$-glycoprotein.

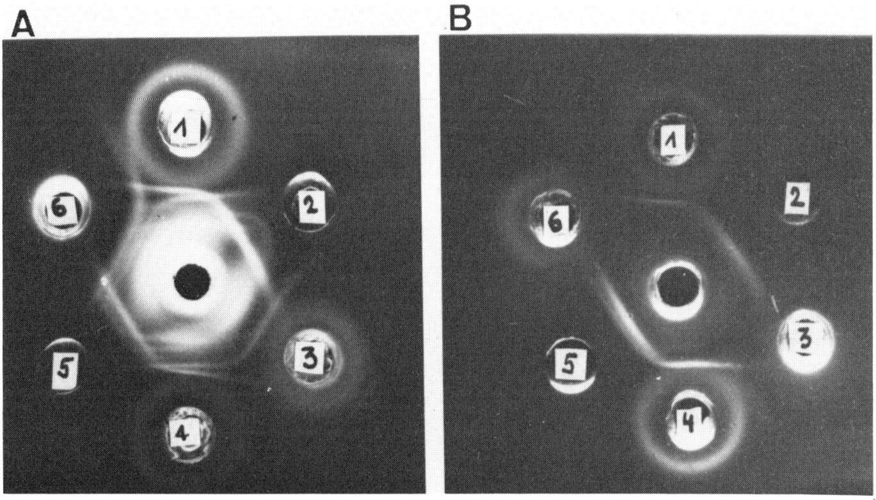

Fig. 17. Immunodiffusion analysis : center well antibrain serum before (A) and after (B) absorption with human serum and liver ; 1 and 4 : brain extract ; 2 and 5 : eluate C_{II} ; 3 : liver extract ; 6 : human serum. Between eluate C_{II} and antibrain serum are seen one (2) or two (5) precipitation lines which show no identity with liver extract or human serum. These lines show identity with brain-specific lines (see also Fig. 19).

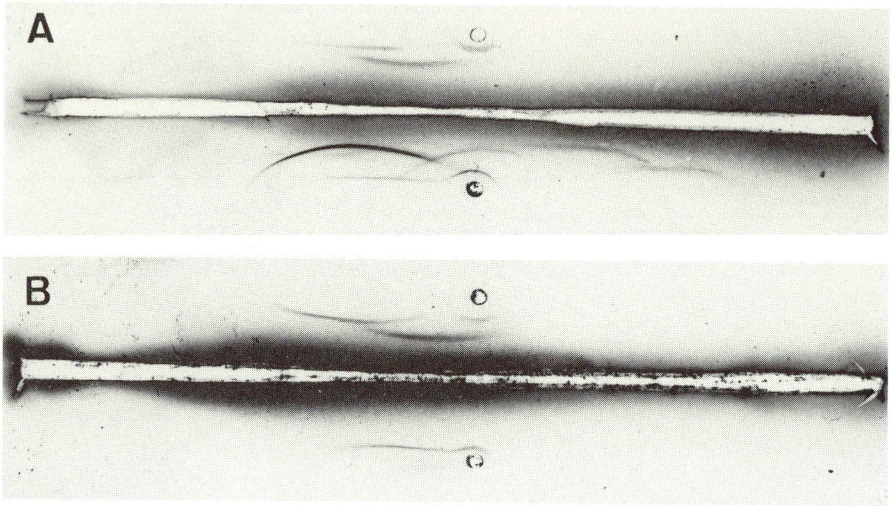

Fig. 18. Immunoelectrophoresis of C_{II} fraction (upper hole A and B) and brain extract (lower hole A and B) with antibrain serum before (A) and after absorption with human serum and liver (B). In the fraction C_{II} two brain specific lines are visible ; one of them corresponds with the alpha$_2$-brain-specific protein, the second protein was enriched by affinity chromatography, it was never seen before using an immunoadsorbent.

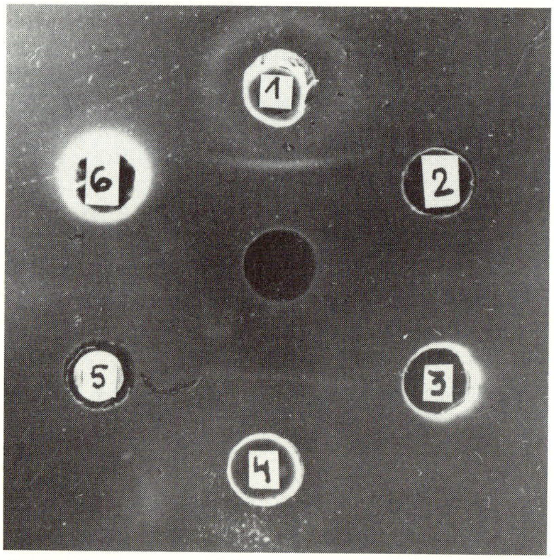

Fig. 19. Immunodiffusion analysis ; after immunization with eluate C_{II} rabbits are building only brain-specific antibodies, no antibodies against liver and serum were detected. Center well:anti-C_{II}-serum, 1 and 4 : brain extract, 2 and 5 : liver extract, 3 and 6 : human serum.

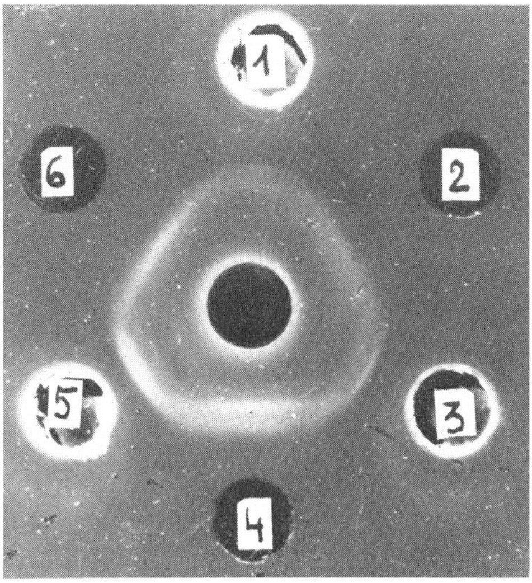

Fig. 20. Immunodiffusion analysis. Center well : anti-C_{II} serum ; 1, 3, 5 : brain extract ; 2, 4, 6 : C_{II} fraction. (See also Fig.19).

Summary

We are dealing with a brain-specific protein bound to a sugar component rich in neuraminic acid. This glycoprotein is a glial protein. It was not found in neurons and in the nuclei of both types of nervous cells. It is localized in microsomes and the supernatant. This protein appears during human ontogenesis at 24-28 weeks of foetal age. This component was almost identical with the brain-specific protein of Macacus rhesus, and semi-identical with the brain-specific protein of rats. This protein can be easily isolated with affinity chromatography. However, by this method, a second brain-specific protein was obtained whose origin is as yet unknown.

ACKNOWLEDGEMENTS

This work was supported by the Deutsche Forschungsgemeinschaft.

REFERENCES

1 AXEN, R., PRATH, J. and ERNBACK, S., Nature, 214 (1967) 1302.
2 GRAY, E.G. and WHITTAKER, V.P., J. Anat., 96 (1962) 79.
3 ROSE, S.P.R., Biochem. J., 102 (1967) 33.
4 TRIPATZIS, I., WARECKA, K. and MAN-CHUNG WONG, Nature, 230 (1971) 250.

5 WARECKA, K., Habilitationsschrift, 1969.
6 WARECKA, K., J. Neurochem., 17 (1970) 829.
7 WARECKA, K. and BAUER, H., Dtsch. Z. Nervenheilk., 189 (1966) 53.
8 WARECKA, K. and BAUER, H., J. Neurochem., 14 (1967) 783.
9 WARECKA, K. and MÜLLER, D., J. Neurol. Sci., 8 (1969) 329.
10 WARECKA, K., MÜLLER, D. and PETERS, K., in preparation.
11 WARECKA, K., MÖLLER, H.J., VOGEL, H.M. and TRIPATZIS, I., J. Neurochem., 19 (1972) 719.
12 WARECKA, K. and VOGEL, H.M., in preparation.

BRAIN GLYCOPROTEIN 10B : FURTHER EVIDENCE OF THE "SIGN-POST" ROLE OF BRAIN GLYCOPROTEINS IN CELL RECOGNITION, ITS CHANGE IN BRAIN TUMOR, AND THE PRESENCE OF A "DISTANCE FACTOR"

Samuel BOGOCH
Dreyfus Medical Foundation, New York ; Boston University School of Medicine and Foundation for Research on the Nervous System, Boston (USA)

Brain glycoprotein 10B was first isolated from normal brain tissue and from brain tumors in 1964 (3). 10B was also isolated from a brain tumor of the ependymoma-glioma type grown subcutaneously in mouse (4), from normal cat hippocampus and caudate nuclei (4), from normal mouse brain, normal pigeon brain, and from subcellular fractions of guinea pig brain (4). 10B is structurally distinguished from microtubule protein (21), S100 protein (18), antigen α (2) and the alkylated brain protein (15). Antisera prepared to brain 10B did not react with proteins of other organs or serum (5).

10B was found to be concentrated six to nine fold in glial brain tumors (4). In addition it was shown to be increased in concentration in Tay-Sach's disease brain in the presence of marked gliosis (9). It was found to be concentrated in glial tumor cells grown in tissue culture (4), and in the outer membranous fractions ("ghosts") of lysates of these glial tumor cells (4). These observations together led to the impression that 10B is a glial constituent. This assumption is clearly supported by the immunochemical data herein reviewed.

10B was shown to increase in concentration in normal pigeon brain during operant conditioning (4,5). ^{14}C-glucose was rapidly and markedly incorporated into the carbohydrate of 10B during training (4, 5, 12). The increase in 10B concentration was followed days later by a return to normal levels of 10B and a concomitant increase in the concentration of the associated brain glycoprotein 11A (4). For this reason, and because of the structural similarities later demonstrated between 10B and 11A, the possibility that 10B is a biosynthetic precursor of 11A has been considered (5).

Sign-Post Theory of Brain Circuitry

I have previously proposed (4, 5) that the glycoproteins of the nervous system are involved in cell recognition, contact, and position functions and that these properties are responsible for the stable inter-cell circuitry necessary for information handling and storage by the nervous system. That is, the glycoproteins determine which nerve cells grow together in contact during development to permit transmission, as well as the facility of transmission at a given synaptic junction throughout life. The experimental evidence in support of this "Sign-Post" theory, summarized in the above references, has included the fact that a change in the structure and function of the brain glycoproteins should accompany both the expression of and the regression of these higher nervous system functions. Evidence for chemical accompaniments of the expression of these functions has been summarized in the above references.The regression of function, as observed in the mental deficiency accompanying certain inborn errors of metabolism, such as Tay-Sachs' disease is associated with glycoprotein pathology (9).

The present report summarizes two types of further evidence in support for the "Sign-Post" theory :

1) Expression. - Changes occur in both the concentration of, and radioactive exchange of the carbohydrates in glycoproteins of the training as compared to resting pigeon brain.

2) Regression. - Changes in the structure of the carbohydrate moieties of the brain glycoproteins occur with the loss of higher cell functions. These changes in glial tumors appear to result in the "unmasking" of certain antigenic sites of 10B, permitting immunofluorescent labelling of their cells of origin by specific 10B antisera. That the unmasking of 10B occurs only in actively dividing astrocytic glia at the growing edge of glial brain tumors suggests that this is an index of malignant change. Malignant change is characterized by the loss of such higher information functions as specific recognition, contact and position of cells. That these changes in brain glycoproteins can be observed in the glycoproteins of the intact brain of mice at some distance from the glial tumor implanted subcutaneously on their backs, as reported here, suggests the existence of a Distance Factor (DF) which may be responsible for the induction of precancerous or early cancerous change in normal cells.

Preparation of 10B

Normal brain or brain tumor tissue is dissected free of meninges and gross blood, and homogenized in 0.005 M phosphate buffer, pH 7, in the ratio of 1 gram of tissue to 100 ml of buffer. For samples under 0.5 g, the Potter-Elvejhem homogenizer is used and for samples greater in size than 0.5 g, the Waring blender is used. Specimens are extracted in the cold room in pre-cooled homogenizers. The first extraction is over a three minute period. The homogenate is centrifuged at 80,000 g for 30 min in a Beckman Model L ultracentrifuge. The supernate is decanted, the residue rehomogenized with a further equal volume of phosphate buffer, centrifuged again, and the second supernate combined with the first. The extraction of the residue is repeated with fresh buffer until less than 50 µg of protein is extracted from the residue. For most specimens this is accomplished in 4 or 5 homogenizations. The soluble proteins thus obtained are combined (extract A), dialyzed at 4° for four hours against 4 liters 0.005 M phosphate buffer, with constant magnetic stirring, then concentrated approximately ten-fold by perevaporation to a final volume of around 15 ml. This material is dialyzed once more against 4 liters of phosphate buffer at 4° with stirring. On completion, the volume of the contents of the dialysis sac is noted, an aliquot is taken for total protein analysis, and the balance is fractionated in the cold room on a DEAE-cellulose (Cellex D) column, 2.5 x 11.0 cm, which has been equilibrated with 0.005 M sodium phosphate buffer. Stepwise eluting solvent changes are made according to the schedule previously given (Ref. 4, p.123) such that solution 1 begins with tube 1, solution 2 with tube 88, solution 3 with tube 98, solution 4 with tube 114, solution 5 with tube 255 and solution 6 with tube 287. Solution 7, which begins with 212, provides groups 10B and 11A, and finally 0.3 percent Triton X-100 is begun with tube 260, to elute groups 11B, 12 and 13. Folin-Lowry protein determinations are made on each tube (4). The total protein in all of the groups thus obtained provides the final yield of protein, in milligram per gram of original weight of brain or tumor tissue. The 10B thus isolated can be rechromatographed on a fresh column of Cellex-D, with the same solvents, with no essential loss.

Properties of 10B

10B differs from microtubule protein (21) in its amino acid composition (see Table I), solubility at acidic pH (3 to 4), and insolubility at neutral and alkaline pH, and in that 10B contains much more carbohydrate (see Table II) than microtubule protein (see Shelanski et al., this Symposium).

10B differs from S-100 protein (18) in several ways :

1) Its amino acid composition is quite different (see Table II);

2) Its carbohydrate content is high, whereas carbohydrate constituents are absent from S-100 ;

3) Immunologically, 10B and S100 are completely distinct (5);

4) S100 appears to be cytoplasmic, whereas 10B appears to be

TABLE I

Amino Acid Composition of Brain Proteins 10B1 and 10B11 (Residues per 1000 residues)

	10B1		10B11	
	Pigeon	Human Tumor	Pigeon	Human Tumor
Glu	129.5	136.8	132.9	139.6
Asp	103.2	102.5	98.7	109.0
Leu	87.7	95.2	85.2	97.9
Ala	87.0	95.7	92.9	99.0
Gly	78.9	73.1	76.8	72.6
Lys	72.2	70.9	80.7	70.3
Ser	66.1	65.4	55.5	65.0
Val	61.4	51.8	63.2	51.0
Thr	52.6	54.8	51.0	54.9
Arg	48.6	51.3	59.4	52.3
Ile	43.8	29.5	43.2	28.3
Pro	41.8	41.7	43.2	47.8
Phe	33.0	36.8	31.8	36.4
Tyr	26.3	26.7	22.6	24.5
His	25.0	26.9	24.5	27.6
Cysteic	23.6	17.1	25.8	10.2
Meth (+cys)	19.6	14.6	12.9	13.3

TABLE II

Carbohydrate Composition of Pigeon Brain Glycoproteins 10B1, 10B11 and 11A, as percent of Protein (Ref.5)

	10B1	10B11	11A
Hexose	11.1	16.7	10.9
Hexosamine	16.2	17.9	4.6
Neuraminic acid	0.6	1.0	0

predominately membranous in localization (4, 19).

10B differs also in its amino acid composition from antigen-α (2).

Incorporation of ^{14}C-Glucose into Training and Resting Pigeon Brain 10B and Other Glycoproteins

Glucose labelled in the first carbon was injected intravenously into pigeons at rest. They were then allowed to rest, or to engage in a training procedure for varying periods of time before sacrifice in a dry ice-acetone bath. The brain glycoproteins were extracted and separated as previously described (4, 5). While this study is more fully described elsewhere (7), Figure 1 summarizes the findings.

At rest for 30 min, only slight incorporation was observed in groups 2, 3 and 11B. Much more incorporation was observed at only 10 min of training.

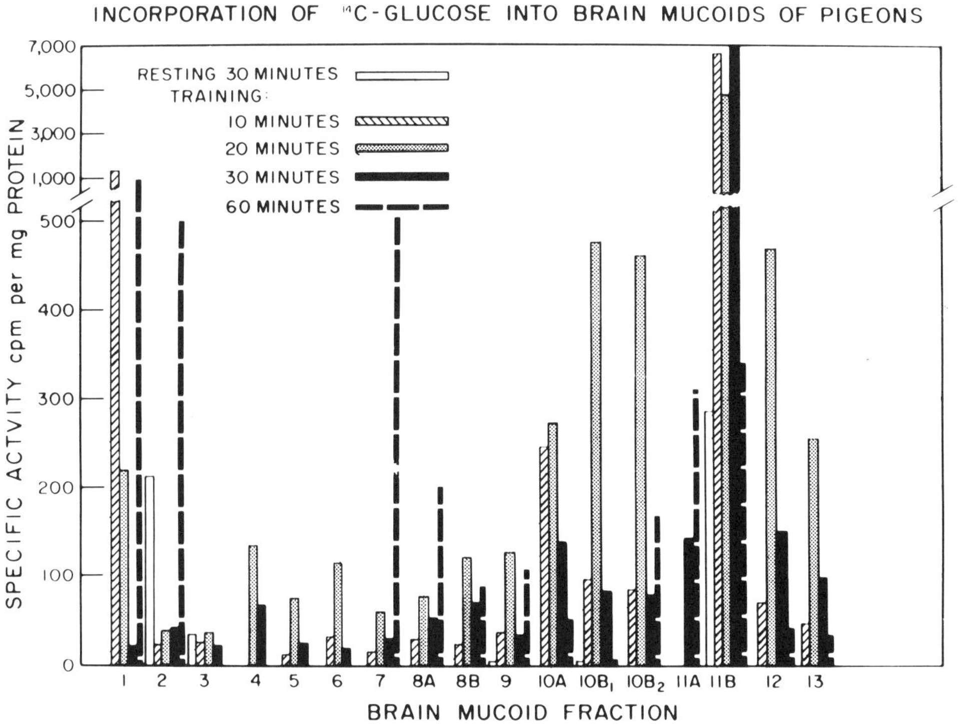

Fig.1

Furthermore, the nature of the most actively labelled groups was a function of time. Thus while at 10 min training, groups 1, 10A and 10B were most prominently labelled; at 20 min, groups 10A, 10B and 10B2, 11B, 12 and 13 were most active; at 60 min, groups 1 and 2 were again active, together with groups 7, 11A and 11B. The fact that groups 1 and 2 are associated with the lateral dendritic processes (4) and that 10B proteins are associated with astrocytic glia, indicate that the sequential activity of glycoproteins of different cellular organelles as well as of different cell typs (glia and neurones) will require further individual examination.

What is clear is that there is an extremely active turnover of carbohydrates in the brain glycoproteins of training pigeons, and that much of this activity occurs within minutes. We had previously noted (4, 5) that these changes occured earlier than those times usually examined in such studies, and the present data confirms the importance of examining discrete and short time periods in relation, where possible, to specific glycoproteins and specific cell fractions.

Concentration of Constituent Hexose of Brain and Tumor 10B and Other Glycoproteins

It was noted earlier that brain glycoprotein 10B concentration is increased in human brain glial tumors and that its protein-bound hexose is reduced (Ref. 4 p.150). In this subsequent analysis of specimens of human brain tumors and of surrounding brain tissue for the concentration and type of glycoprotein present, the total protein-bound hexose in the tumor itself, 2.7 percent, was less than that in surrounding glial capsule, 6.2 percent, or in the outside normal cortical tissue, 5.1 percent.

In cases where the demarcation of tumor from surrounding tissue was not as clear as that above, and the degree of infiltration of normal tissue by tumor cells was variable, it was difficult to impossible to carry out this type of comparative analysis. Edema and hemorrage in the surrounding tissue also militate against such study.

To avoid this problem, ependymoma-glioma tumors previously grown in C3H mice, were transferred to subcutaneous sites on the back of the mouse, in doses of 5 to 20 mg/mouse, to permit analysis of the tumor as well as the brain of the same mouse, free of local pathology. The tumors were grown for two weeks, the mice being kept in individual cages to avoid canabalism and to keep track of the volume of fluid intake. The mice received diphenylhydantoin (DPH), or its solvent alone ; in their drinking water, of which the volume consumed daily was recorded. DPH has been found to increase brain free glucose and glycogen (13, 16, 22) and to

TABLE II

Total Hexose Bound to all Protein Groups of Mouse Brain, as Percent of Protein

	No DPH	With DPH
Brain, control	4.6 , 5.5	7.9 , 8.0
Brain, in "Tumor-bearer"	2.6 , 2.3	6.8 , 7.0

TABLE IV

Hexose Bound in Glycoprotein 10B of Mouse Brain, as Percent of Protein

	No DPH	With DPH
Brain, control	5.4	4.8
Brain, in "Tumor-bearer"	3.7	10.5

counteract anoxia (see Ref. 11). The mice were sacrificed, and tumors and brains from groups of six mice were each pooled and extracted for total glycoproteins.

In tumor-bearing animals, total brain protein-bound hexose (Table III) and hexose in isolated brain glycoprotein 10B (Table IV) were observed to be greater in concentration in DPH-treated than in non-treated mice. In normal mouse brain, (non tumor-bearing) the concentration of total brain protein-bound hexose has not constantly been observed to increase with DPH, but under certain conditions to decrease or to remain unchanged (10).

Stepwise Hydrolysis and Identification of Constituent Carbohydrates of Glial Tumor Glycoproteins

The total glycoproteins were hydrolysed by means of a stepwise acidic procedure previously described (5). The liberated sugar residues were isolated by dialysis at each hydrolytic step and separated from the amino acids and salts by means of a two phase column chromatography procedure (10). The separated sugars and amino acids were then studied by thin-layer chromatography as earlier described (4).

TABLE V

Constituent Carbohydrates of the Isolated Glycoproteins of Pigeon Brain and Glial Tumor

	Pigeon Brain			Glial Tumor	
	10B1	10B11	11A	No DPH	With DPH
Rhamnose (?)	+	+			+
Fucose (ribose)	+	+		+	+
Xylose	+	+			
Mannose	+	+	+	+	+
Glucose	+	+	+	+	+
Galactose	+	+	+	+	+
Mannosamine (?)	+	+			+
Glucosamine		+		+	+
Galactosamine	+	+		+	+
Oligosaccharide 1	+				+
Oligosaccharise 2	+				+

The results are summarized in Table V. There were several differences :

1) A fast-running sugar with Rf of rhamnose was observed only in the DPH-treated tumor glycoproteins.

2) An amino sugar with the Rf of mannosamine was prominently released in the DPH-treated tumor but was questionable or absent in the non-DPH treated tumor.

3) Two sugars with the mobility of oligosaccharides, labelled oligosaccharide 1 and 2 in Table V, were released prominently in the DPH-treated group, but were absent in the non-DPH treated tumor.

4) In DPH-treated brain, the glycoproteins demonstrated the preponderance of galactose ; whereas galactose did not appear to be present,or was present in trace amounts only, as a readily released sugar in non-DPH treated material.

Subtler differences observed which may merit further consideration in future studies are as follows : with DPH treatment,galactosamine appears to be preponderant over glucosamine, whereas in the non-DPH treated tumors the reverse holds, that is, glucosamine appears to be preponderant over galactosamine. These differences in sugars as demonstrated by thin-layer chromatography are now being investigated by gas chromatographic methods.

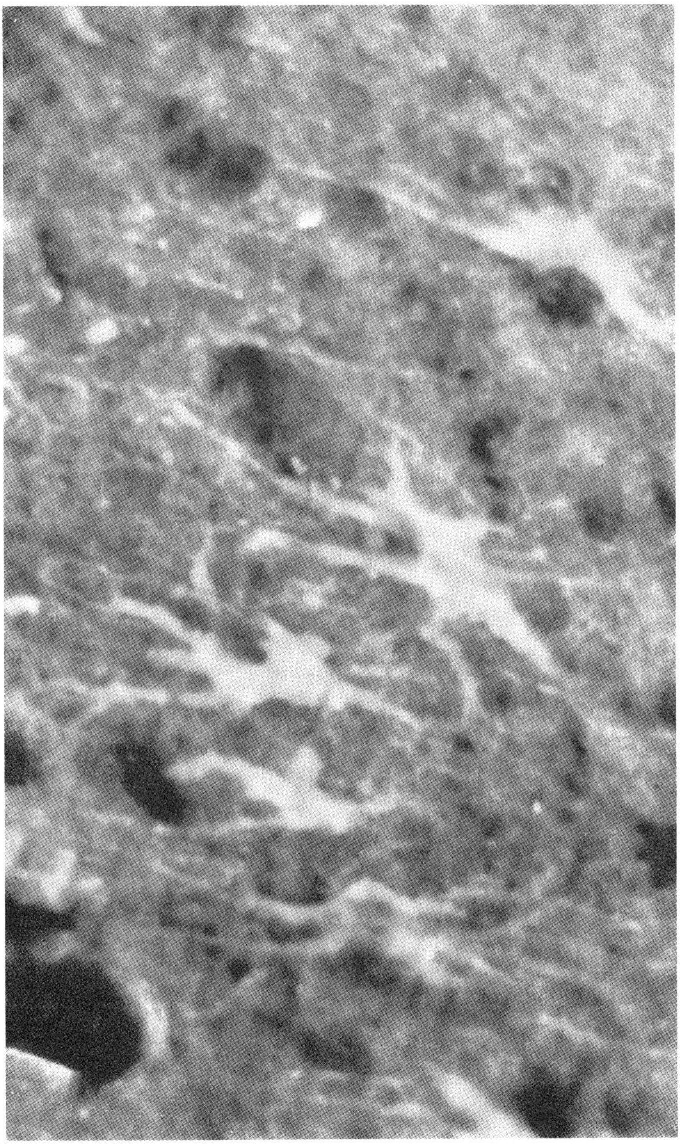

Fig. 2. Immunofluorescent staining of reactive astrocytes in glioblastoma ; produced with antiserum to brain glycoprotein 10B (Ref. 1).

Immunological Studies of 10B : Specificity for Reactive Astrocytes

Utilizing an antiserum to 10B, prepared in rabbits, it has been possible to show that, at highest dilutions, immunofluorescence occurs only in the astrocytic glia, and exclusively in "reactive" astrocytes, both in astrocytomas and in glioblastoma multiforme (1) (Fig.2). No staining of frankly neoplastic cells, of normal glia, or of ganglion cells occured.

These observations are in agreement with earlier data, cited in the introduction, indicating that 10B is a glial constituent. That it is accessible to fluorescent staining, that is, to combination with specific antiserum in situ only when in a "reactive" or dividing state suggests that either its structure is altered in this state, and/or that it has become "unmasked" by the removal of some blocking cell constituent or group, which would normally not permit applied 10B antibody to contact the 10B antigen and react with it. The implications of such a change in cell constituent reactivity in relation to the development of the malignant process will be taken up in the discussion.

From a practical point of view, both with regard to diagnostic and therapeutic potential, it became of interest to determine how specific the antiserum is to a particular 10B which is isolated from a given human or animal brain tumor. To this end we have begun to extract and purify 10B from each specimen of human brain tumor, and from each specimen of surrounding more or less normal brain tissues, and separately to prepare antisera to these 10B antigens (10).

After concentration of 10B antigen in solution at 0.1 - 0.2 mg/ml there is some precipitation of 10B. Only the soluble supernate is injected, after Folin-Lowry protein determination. The first injection is with complete Freund's adjuvant, into the foot pads ; intravenous booster is given 10 days later ; intramuscular injection in complete Freund's adjuvant is given two weeks later. Rabbit sera were tested on Ouchterlony double-diffusion for the degree of antibody formation.

Table VI records the data to date. Each reactive line is denoted a or b, and its intensity graded + to +++. It may be seen that without cross-absorption or other purification of either antisera or antigens, there is some indication of specificity and some of cross-reactivity. Cross absorption and other purification steps are now in progress for both antigen and antibody. It is important to note that the ability of isolated antigen and antibody to combine in a gel or in solution without impediment is quite a different matter from reaction of antibody with in situ cellular 10B antigen. Thus 10B antigen is available in reactive astrocytic glia to

TABLE VI

Cross Reactions between 10B-Antigens and Their Antisera Isolated :
a) from Human Glial Tumors and
b) from Surrounding "Normal" Brain Tissue

	Antibody to :			
	I.J."Normal"	I.J.Tumor	J.C."Normal"	J.C.Tumor
10B Antigen from:				
I.J. "Normal"	a+	a+	-	-
I.J. Tumor	a++	a+++	a++ b++	-
J.C. "Normal"	-	a++	a+++ b+	a++
J.C. Tumor	-	a+	a+++	a+++

combination with 10B antibody, but it is unavailable for such combination in non-reactive astrocytes.

Discussion

Search for the chemical expression of higher information functions of brain cells has led in addition to findings of structural changes in the same molecules upon regression of these higher functions, in brain tumors. The antisera which we have obtained to a brain tumor cell antigen 10B have possible application to both diagnosis and therapy of brain tumors. Thus since only the 10B of reactive astrocytes will combine with added 10B antisera, it may be possible to outline radioactively a brain tumor for active viewing on screens, or photography for permanent record. Further, if the 10B antibody injected is coupled with chemically or radioactively toxic compounds, or with such compounds which can be activated by chemical or physical means, destruction of cells in the brain tumor area can be much more selective in sparing normal cells than hitherto possible. In addition, if circulating antibody to brain tumor antigen 10B can be detected in human serum, a further diagnostic procedure may become available.

The growth of ependymoma-glioma tumors subcutaneoulsy in the mouse provides the opportunity to study a brain tumor growing outside the nervous system, and at the same time to study the chemistry of the essentially normal brain (and other organs) in the same animal able to be influenced by the subcutaneously growing brain tumor. The decrease in protein-bound hexose of brain under these conditions suggests the influence of a diffusable factor from the distant brain tumor. As pointed out in the introduction, this factor may be responsible for the induction of a precancerous or early cancerous change in normal cells.

The concentration of total protein-bound hexose in brain under the influence of distant tumor is in agreement with the data obtained for human brain tumors. It is also in general agreement with the previous observations that the concentration of protein-bound hexose in brain is greater : (a) the more complex the anatomical structure of the brain, (b) the more ordered the chemical structure, as in membranes opposed to cytoplasmic cell constituents, and (c) the more complex or active the functional state of the animal (4,5). Thus, the greater the "experiential" or environmental informational content or activity of a given structure or function, the greater the concentration of protein-bound hexose. The training brain cell, with relatively high experiential informational content, contains more protein-bound hexose (4). The tumor cell, with relatively low experiential informational content, but high "genetic" activity, contains less protein-bound hexose.

This relationship may be generalized to the proposition that I have made that when DNA and cell replication functions are stimulated, as in tumors, normal mucoid biosynthesis is inhibited (8).

During normal morphogenesis of brain, DNA and cell division would be more active at one stage, and mucoid biosynthesis for specific interneuronal connections would be more active at another stage (4).

This postulated inverse relationship of mucoid-contact functions to DNA replication could account for the cessation of DNA synthesis in retinal ganglion cells observed to be correlated with the time of specification of their central connections (17). The theory would also be consistant with the reported relationship of malignancy to loss of contact inhibition (20), if contact recognition is indeed a function of mucoids. This cell surface difference in malignancy has been related to the presence of an agglutinin with specificity for the N-acetylglucosamine determinant (14). If nucleic acid bases, uridine and cytidine, were available either for nucleic acid synthesis or for transferase activity for mucoid synthesis, but not for both simultaneously, then a control mechanism would be at hand for determining which of the two cell functions (replication or contact positioning) occured.

Summary

1. The isolation and properties of brain glycoprotein 10B are described. 10B is increased in concentration in glial tumors in humans and animals, but its hexose content is reduced.

2. Specific antibodies to 10B have been prepared. The in situ 10B antigen is only available for staining by antibody immunofluorescence when the astrocytic glia are in the "reactive" state as when around glial brain tumors.

3. When ependymoma-glioma tumors are grown subcutaneously in the mouse at a distance from the brain of the animal, changes in the brain glycoproteins can be detected, suggesting the presence of a Distance Factor (DF) which may be responsible for the induction of cancerous changes in distant cells. Individual 10B antigens are now being isolated from individual human brain tumor and surrounding tissue specimens, and antisera prepared to test the diagnostic and therapeutic potential of the availability of specific antibody to this specific brain tumor cell antigen.

4. 10B concentration is increased during one phase of training in the pigeon. ^{14}C-glucose is incorporated more rapidly into 10B during training than during resting. During one hour of training, 10B and other brain glycoproteins are labelled at different times.

5. Carbohydrate moieties of glycoproteins of brain cells and tumor cells have been isolated and compared by thin-layer chromatography.

6. The "Sign-Post" theory (4) of brain glycoprotein function is supported. In addition, recent evidence has led to the extension of the concept to the proposal that mucoid biosynthesis is inversely related to DNA replication. That is, DNA and cell division are inhibited during cell positioning and the formation of inter-cell contacts (the proposed glycoprotein Sign-Post "experiential" function), and mucoid biosynthesis is inhibited when DNA and cell division (the genetic function) is active.

REFERENCES

1 BENDA, P., MORI, T. and SWEET, W.H., J. Neurosurg., 33 (1970) 281.

2 BENNETT, G.S. and EDELMAN, G.M., J. Biol. Chem., 243 (1968) 6234.

3 BOGOCH, S., in Eight Internatl. Congr. Neurol., Internatl.Congr. Ser., (Ed. Constans, J.P., Gibberd, F.B. and Junze, K.) Excerpta Medical Foundation, Amsterdam, 1965, p.139.

4 BOGOCH, S., The Biochemistry of Memory ; With an Inquiry into the Function of Brain Mucoids, Oxford University Press London, 1968.

5 BOGOCH, S., in Protein Metabolism of the Nervous System (Ed. Lajtha, A.) Plenum Press, New York, 1969, p.555.

6 BOGOCH, S., in Handbook of Neurochemistry (Ed. Lajtha, A), Plenum Press, New York, 1969, vol. 1, p.75.

7 BOGOCH, S., in Proc. Internatl. Conf. on Sphingolipidosis and Allied Disorders, New York, 1971 (Ed. Aronson, S.M. and Volk, B.W.) in press.
8 BOGOCH, S., Third Internatl. Meeting Internatl.Soc. Neurochem., Budapest, 1971, Abstr.
9 BOGOCH, S. and BELVAL, P., in Inborn Disorders of Sphingolipid Metabolism. Proc. 3rd Internatl. Symp. Cerebral Sphingolipidoses (Ed. Aronson, S. and Volk, B.W.) Pergamon Press,New York,1966, 273.
10 BOGOCH, S., BOGOCH, E., KORSH, G. and DAS, B.R., unpublished results.
11 BOGOCH, S. and DREYFUS, J.J., in The Broad Range of Use of Diphenylhydantoin : Bibliography and Review. Dreyfus Medical Foundation, New York, 1970.
12 BOGOCH, S., SACKS, W., SWEET, W.H., KORSH, G. and BELVAL, P.C., in Protides of the Biological Fluids, Proc. 15th Coll.,Bruges, 1966 (Ed. Peeters, H.) Elsevier Publishing Co., New York,1968, p.129.
13 BRODDLE, W.D. and NELSON, S.R., Feder. Proc., 27 (1968) 751.
14 BURGER, M.M. and GOLDBERG, A.R., Proc. Natl. Acad. Sci. US, 57 (1967) 359.
15 FALXA, M.L. and GILL, T.J., Arch. Biochem. Biophys., 135 (1969) 194.
16 HUTCHINS, D.A. and ROGERS, K.J., Brit. J. Pharmacol., 39 (1970) 9.
17 JACOBSON, M., Develop. Biol., 17 (1968) 219.
18 MOORE, B.W. and McGREGOR, D., J. Biol. Chem., 240 (1965) 1647.
19 QUAMINA, A. and BOGOCH, S., in Protides of the Biological Fluids, Proc. 13th Coll., Bruges, 1965, (Ed. Peeters, H.) Elsevier Publishing Co., New York, 1966, p.211.
20 STOKER, M. and RUBIN, H., Nature (Lond.) 215 (1967) 171.
21 WEISENBERG, R.C., BORISY, G.G. and TAYLOR, E.W., Biochemistry, 1 (1968) 4466.
22 WOODBURY, D.M., TIMIRAS, P.S. and VERNADAKIS, A., in Hormones, Brain Function and Behavior (Ed. Hoagland, H.) Academic Press, New York, 1957, p.38.

SECTION 2

PROTEINS OF FUNCTIONAL IMPORTANCE IN THE NERVOUS SYSTEM

SOME BIOCHEMICAL ASPECTS OF NEUROTUBULE AND NEUROFILAMENT PROTEINS

Michael L. SHELANSKI, Howard FEIT, Robert W. BERRY and Matthew P. DANIELS

Department of Pathology (Neuropathology), Albert Einstein College of Medicine, Bronx, New York 10461 (USA)

In the few years since the isolation of the subunit protein of microtubules (1, 2, 23, 26) work on its chemistry, localization and metabolism has proceeded at a feverish pace. In their initial studies Taylor and his co-workers noted exceptionally high levels of colchicine-binding activity in brain homogenates and in axoplasm. Weisenberg, Borisy and Taylor (31) developed a simple procedure for the purification of the microtubule subunits (tubulin) from brain enabling investigators to use large quantities of highly purified protein for biochemical studies. Studies on metabolism and distribution of tubulin in brain were aided by the observation that vinblastine quantitatively and selectively precipitated tubulin from 100,000 g supernatants of brain homogenates (17, 18, 32).

Colchicine binding and vinblastine precipitation techniques have been utilized in a number of laboratories to study tubulin and the role of microtubules in neuronal function. Since this work has been extensively reviewed elsewhere (25) we shall confine ourselves to a few newer and provocative observations from our own laboratory.

Before presenting our observations it is useful to review briefly some chemical properties of tubulin. Tubulin from brain, like tubulin from a number of other cytoplasmic sources, is a dimer with a sedimentation velocity of 6S and is composed of monomers of molecular weight of approximately 55,000 each. On disc gel electrophoresis the monomers may be separated and amino acid compositions (3) and peptide maps (9) of the two species show small dissimilarities. It is most tempting to interpret this data as showing non-identity of tubulin monomers but the possibility of two distinct tubulin species in brain has not been ruled out (19). Tubulin is capable of binding one mole of colchicine per dimer (2) and has one mole of tightly bound and one mole of readily exchangable guanine nucleotide per dimer (31). Perhaps the most striking thing about tubulin in brain is the quantities which are present and the rapidity with which it is turned over. Tubulin accounts for at least 15 percent of the soluble protein of mouse brain (6) and 28 percent of the soluble protein of isolated synaptosomes (8). Tryptic peptide maps show that tubulin obtained from synaptosomes is similar if not identical to tubulin from the soluble supernatant of whole brain. After a single intracerebral injection of leucine-^{3}H, tubulin turns over with a half-life of 4-5 days.

In view of the large quantity of tubulin present in brain and its rapid turnover, one is curious about its potential function. One function often proposed is cytoskeletal. To be sure, neurons must preserve great asymmetry and a skeleton of microtubules could be of use. However, it seems unlikely that so much rapidly turning over protein would be required for this function alone. A second function often proposed is some part of the motile mechanism in axoplasmic transport. Numerous studies have demonstrated (for review see Ref. 25) blockade of axoplasmic flow by the drugs colchicine and vinblastine. Other studies have shown the transport of tubulin itself as a major component of the slowly transported phase (8, 28).

Side arms on axonal microtubules (10) and cross bridges between microtubules and synaptic vesicles (14) and the association between tubules and mitochondria in the axon (21) have also been reported. One possible mechanism would have the "arms" on the tubule function as the motile element moving particles and organelles down the axon. The actual role of microtubules in axoplasmic flow remains a mystery.

Phosphorylation and Glycosylation

The recent observations that endocrine release functions from pancreas (15), adrenal (20) and thyroid (33) were blocked by colchicine and vinblastine, led Rasmussen and his associates to an investigation of cAMP stimulated phosphorylation of tubulin as a possible step in this mechanism (11). Their observation of cAMP stimulated, kinase-dependent phosphorylation of tubulin *in vitro* is interesting but has not been demonstrable *in vivo* (6).

A second type of modification reaction which might play a role in secretory events is glycosylation. Feit has demonstrated incorporation of glucosamine-^{3}H into the protein as both glucosamine and galactosamine. Reducing sugars have been reported as being present at concentrations of 1-2 percent (5, 11). Total hexosamine concentration is about 1 percent and fucose, glucose, galactose, and

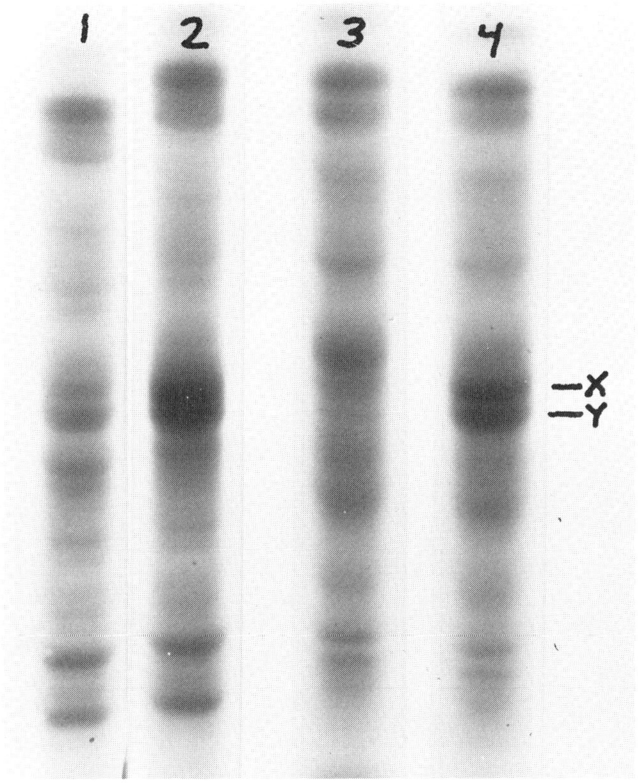

Fig. 1. Comparison of the electrophoretic pattern of nerve-ending ghosts (1) and synaptic vesicles (3). Discontinuous urea-SDS polyacrylamide gel electrophoresis. Nerve endings were isolated from 34 adult mice, homogenized in water with a Dounce homogenizer fitted with a tight pestle, frozen, thawed and rehomogenized. The homogenized nerve endings were centrifuged at 100,000 g for 1 h and soluble components were removed. The particulate fraction was resuspended in water and centrifuged through a discontinuous sucrose gradient that contained equal volumes of 0.4 M and 1.2 M sucrose. The fractions that appeared on the 0.0 M - 0.4 M interface (vesicles) and on the 0.4 M - 1.2 M interface (ghosts) were used for these gels. The gels were stained with amido schwartz. The positions of the x and y subunits of tubulin are indicated. (1) Nerve-ending ghosts. (2) Ghosts to which WS-tubulin was added. (3) Synaptic vesicles. (4) Synaptic vesicles to which WS-tubulin was added.

sialic acid are associated with the protein (Margolis, Margolis and Shelanski, unpublished observations).

Tubulin has a limited ability to hydrolyze GTP with the liberation of Pi. Preliminary studies indicate that 2-3 moles of Pi is liberated per mole protein per hour at 37° in the presence of excess GTP. The mechanism or specificity of this reaction remains to be established, however, it is interesting that GMPPCP, the non-hydrolyzable analog of GTP,has been observed to inhibit vinblastine precipitation of tubulin and to stabilize colchicine binding activity (Ventilla, M., personal communication).

Tubulin and Insoluble Colchicine-Binding Activity

While over 90 percent of the colchicine binding activity in KB cells is soluble (1) and similar distribution of binding activity is seen in other cells and tissues, brain is an apparent exception. In brain almost half of the colchicine binding activity is particulate (7). This activity is localized to microsomal and synaptosomal membrane fractions (7, 16). The characteristics of the membrane associated binding site were similar to those of soluble tubulin but attempts to solubilize the membrane receptor with colchicine-binding activity intact failed because all treatments which released appreciable amounts of protein from the membrane destroyed both membrane-associated and soluble colchicine binding activity. Thus, attempts were made to visualize protein in the membrane which co-migrated with soluble tubulin on gel electrophoresis. It was found that in synaptosomal membranes there was a tight double band of protein which co-migrated exactly with the doublet from purified tubulin (8). Furthermore, comparison of membrane fractions having colchicine binding activity with those lacking colchicine-binding activity showed the tubulin doublet in the former but not in the latter.

In contrast to the membrane fractions, synaptic vesicles showed only very faint bands co-migrating with tubulin (Fig. 1) and these became fainter as the purity of vesicle fractions increased.

Amino Acid Composition after Preparative Gel Electrophoresis

Samples of WS-tubulin and NEP from 3-7 day-old mice were run on the discontinuous SDS-urea gel system and the corresponding x and y subunit bands were cut out of fixed, stained (amido black) gels. The protein was eluted from the gel slices by an electrophoretic technique (9) using 0.1 percent SDS in 0.1 M phosphate buffer pH 7.0. The eluted protein was concentrated by vacuum dialysis and at least 50 volumes of water were passed through each protein solution to remove some, but not all, of the SDS and residual stain.

The dialyzed protein solutions were diluted with an equal volume of concentrated HCl and hydrolyzed at 110° for 24 hours in sealed, evacuated tubes. After hydrolysis, the samples were dried under a stream of nitrogen. The amino acid composition of each sample, expressed as moles percent recovered, is shown in Table I. The compositions agree in a general way except for the low values for threonine and serine in the presumptive tubulin subunits from the NEP (nerve ending particulate) material. If the values for threonine and serine in the WS-tubulin subunits are used to recalculate the composition of NEP_x and NEP_y, the agreement between the amino acid compositions of the corresponding subunits improves. This recalculation is probably justified because a new peak was present in the NEP_x and NEP_y analysis that was not present in the WS-tubulin x and y analysis (Fig. 2). This peak elutes in a position that would correspond to an acidic derivative of either threonine or serine, perhaps pyruvate. However, no direct evidence for a relation between the new peak and the loss of threonine and serine is available.

TABLE I

Amino Acid Compositions after Preparative Gel Electrophoresis
(In moles % recovered)

	Tubulin subunit x	NEP x	Recal-culated NEP_x	Tubulin subunit y	NEP y	Recal-culated NEP_y
Lys	4.40	5.20	9.41	4.75	5.41	8.62
His	2.32	2.74	2.53	2.15	2.43	2.22
Arg	4.21	4.00	4.23	4.80	4.82	4.99
Asp	8.63	9.47	8.73	9.42	9.21	8.44
Thr[1]	2.49	0.66	(2.49)	4.09	0.36	(4.10)
Ser	2.84	0.80	(2.84)	5.92	0.52	(5.92)
Glu	11.76	12.50	11.53	12.02	12.09	11.08
Pro	4.14	4.28	3.95	4.30	4.68	4.28
Gly	14.35+	14.76+	13.61+	13.53+	12.90+	11.82+
Ala	7.31	8.21	7.57	6.64	7.94	7.27
Cys[2]	-	-	-	-	-	-
Val	8.37	8.97	8.27	7.56	8.25	7.56
Met	1.19	1.69	1.56	1.60	1.75	1.60
Iso	4.61	5.04	4.65	4.51	4.93	4.51
Leu	8.07	8.02	7.40	8.29	9.05	8.29
Tyr	0.30	0	0	0	0	0
Phe	3.98	2.60	2.40	3.78	4.08	3.78

+Not reliable, may contain some glycine from electrophoresis buffers. [1]Thr and Ser not corrected for hydrolytic destruction. [2]Carboxymethyl cysteine not determined.

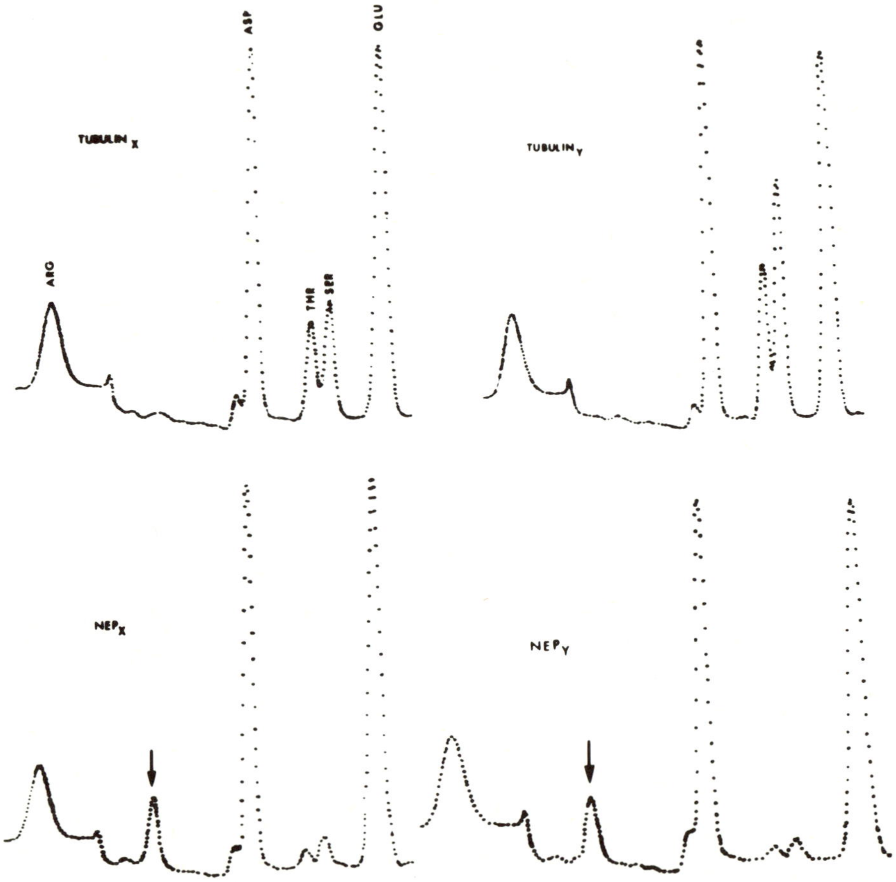

Fig. 2. Amino acid analysis of tubulin subunits. The x and y subunits of WS-tubulin and the corresponding subunits in the NEP fractions were isolated by preparative polyacrylamide gel electrophoresis and analyzed on a Jeolco 5AH amino acid analyzer with a two-column methodology. A portion of the chromatogram for each subunit has been traced to qualitatively illustrate the loss of threonine and serine residues and the appearance of a new compound (arrow) in NEP_x and NEP_y. The new compound appeared 40 min after the start of the elution of the long column.

This experiment suggests that in becoming associated with the membrane, tubulin undergoes extensive modification of its threonine and serine residues. Two such well-known reactions of serine or threonine are glycosylation or phosphorylation. Serine or threonine residues with O-glycosidic bonds undergo a β-elimination reaction under alkaline conditions to form the corresponding α-β-unsaturated amino acid (12). Our samples were reduced and alkylated at pH 8.8. However, the elimination reaction is probably not specific for glycosidic leaving groups. Studies with model compounds have employed a variety of leaving groups, including di-O-phenylphosphoryl and toluene-p-sulphonyl (12). After β-elimination, the corresponding α-β-unsaturated amino acid residues are very labile to either mild acid or base, and the double bond could undergo addition reactions. Such residues may be converted by treatment with sodium borohydride to alanine and α-aminobutyric acid (12). If after treatment of the NEP material with sodium borohydride, there were an increase in alanine and the appearance of α-aminobutyric acid in the amino acid composition, this would constitute additional evidence for modification of threonine and serine residues. However, we found that after treatment with 0.1 N NaOH and 0.3 M sodium borohydride, tubulin was extensively degraded. This precluded the possibility of doing the experiment by treating the particulate fraction with sodium borohydride before isolating the presumptive tubulin.

Current attempts to isolate the presumptive tubulin from nerve ending ghosts and compare it to soluble tubulin by cyanogen bromide peptide mapping methods may shed more light on these observations.

Solubilization Experiments

Although conditions for the solubilization of particulate colchicine-binding activity have not been found, the material presented below suggests that this activity was not due to adsorption or entrapment of soluble tubulin. Two experimental approaches were used. In some experiments, solubilization of the colchicine-binding material in washed, whole particulate fractions of brain was attempted using colchicine-binding as the assay. The criterion for solubilization used in these experiments was the loss of particulate activity with the concomitant appearance of soluble activity. In other experiments, the NEP fraction was exposed to various conditions (which might destroy colchicine-binding activity) and the solubilized (100,000 *g* supernatant) material and the remaining material were analysed by gel electrophoresis on the discontinuous SDS-urea gel system.

Ionic strength. - Increased ionic strength reduces the effectiveness of ionic binding. Particulate colchicine-binding activity was not extracted by 0.6 KCl nor was it lost from the particulate fraction. 2 M KCl removed only a small fraction of the material with the same electrophoretic mobility as tubulin in the NEP fraction.

Detergents. - A variety of detergents were screened for their effect on the colchicine-binding assay. The whole soluble fraction was made 0.1 % in either Brij 35, NP40, Triton X-100, Sarkosyl, digitonin, Tween, SDS, or sodium deoxycholate, and colchicine-binding activity was measured. Triton, Tween and sarkosyl caused less than a 20 percent decrease in colchicine binding. However, when a homogenate was made and aliquots of this homogenate were removed and rehomogenized in the presence of 0.1 percent Triton, Tween or sarkosyl, colchicine-binding activity was diminished by at least 70 percent in both the supernatant and particulate fraction when compared to non-detergent treated material. It is possible that these detergents released a factor, proteolytic or otherwise, which destroyed colchicine-binding activity. If washed particulate fractions were extracted with 0.1 percent Triton, a factor with colchicine-binding activity was released, but there was no concomitant decrease in the specific activity of the remaining particulate material. Moreover, if the colchicine was first incubated with the particulate material before treatment with Triton, the detergent did not extract the bound colchicine. Triton removed only a small fraction of the material with the same electrophoretic mobility as tubulin from the NEP fraction.

EDTA. - This agent has been used to solubilize the central pair of microtubules of sea urchin sperm tails (26). After soluble and particulate fractions were dialyzed against 1 mM-EDTA at 4° for 1 h, the fractions retained 45 percent and 34 percent respectively, of their initial colchicine-binding activity. If these fractions were then dialyzed against 20 mM-$MgCl_2$ for 1 h at 4°, the soluble material had 55 percent of its initial activity, and 98 percent of the original particulate binding activity was recovered in the high speed pellet from the dialyzed particulate material ; no soluble colchicine-binding factor was released. EDTA was also not very effective in removing the presumptive tubulin from the NEP fraction.

Trypsin. - Treatment with low concentrations of trypsin is believed to remove protein from the surface of membranes. Treatment of the NEP fraction with trypsin (30 microgram /ml, 2 h, 30°) failed to solubilize the material which had the same electrophoretic mobility as tubulin in this fraction.

Disulfide bonds. - The possibility that the tubulin was associated with the membrane through disulfide bonds was considered. Treatment of the NEP fraction with 0.1 M mercaptoethanol for 2 h at room temperature was no better than treatment with water in removing the presumptive tubulin from the NEP fraction.

pH. - In sea urchin sperm tails, the outer doublet microtubules are composed of two closely related proteins with very different solubility characteristics (29). Thus, B-tubulin is soluble at neutral pH whereas A-tubulin is almost insoluble at this pH. Although it is not known if brain has species of tubulin comparable to A- and B-tubulin, this possibility was considered. Above pH 8.0, colchicine-binding activity was irreversibly destroyed. Wilson (34) also observed irreversible destruction of colchicine-binding activity above pH 8.0. It thus could not be ascertained, using the colchicine-binding assay, if the particulate activity corresponds to a tubulin with the solubility characteristics of A-tubulin. However, strong base (at least 0.025 N NaOH, 0°, overnight) was very effective in solubilizing the NEP fraction. An attempt is being made to isolate tubulin from the base-treated NEP fraction. The fact that base dissociates this membrane fraction is consistent with the other evidence for the involvement of serine or threonine residues in O-glycosidic or similar bonds when protein is associated with the membrane.

The solubilization experiments suggested a "tight" association of the colchicine-binding material with the particulate site. Interfering with ionic, hydrophobic, disulfide, or divalent cation-stabilized bonds was in each case not a sufficient condition for the effective release of tubulin or tubulin-like material from its particulate site. These experiments made unlikely the possibility that particulate colchicine-binding activity was caused by non-specific adsorption of tubulin. Further evidence against this possibility was the failure to detect significant cross-contamination between the soluble and particulate fractions when particulate fractions were prepared from homogenates that contained previously labelled soluble protein. Also, testis and neuroblastoma cells each had high levels of soluble colchicine-binding activity and low levels of particulate colchicine-binding activity. In brain, the particulate activity was associated specifically with certain membrane components and not others. And finally, amino acid analysis of the proteins from the NEP fraction which presumably were related to the colchicine-binding activity of this fraction, suggested that these proteins were not identical with tubulin but were possibly derived from it through modification of threonine and serine residues. Final clarification of the relation between soluble and particulate colchicine-binding activity awaits a suitable peptide mapping experiment.

The results on particulate colchicine binding presented above are provocative but by no means conclusive. One might imagine synthesis of tubulin in the cell soma, transport to the ending and membrane-association as having some role in neurosecretory events. Rasmussen (22) has proposed such a schema involving both tubules and a contractile releasing mechanism. His specific proposal of a tubulin-synaptic vesicle complex is not supported by our observations

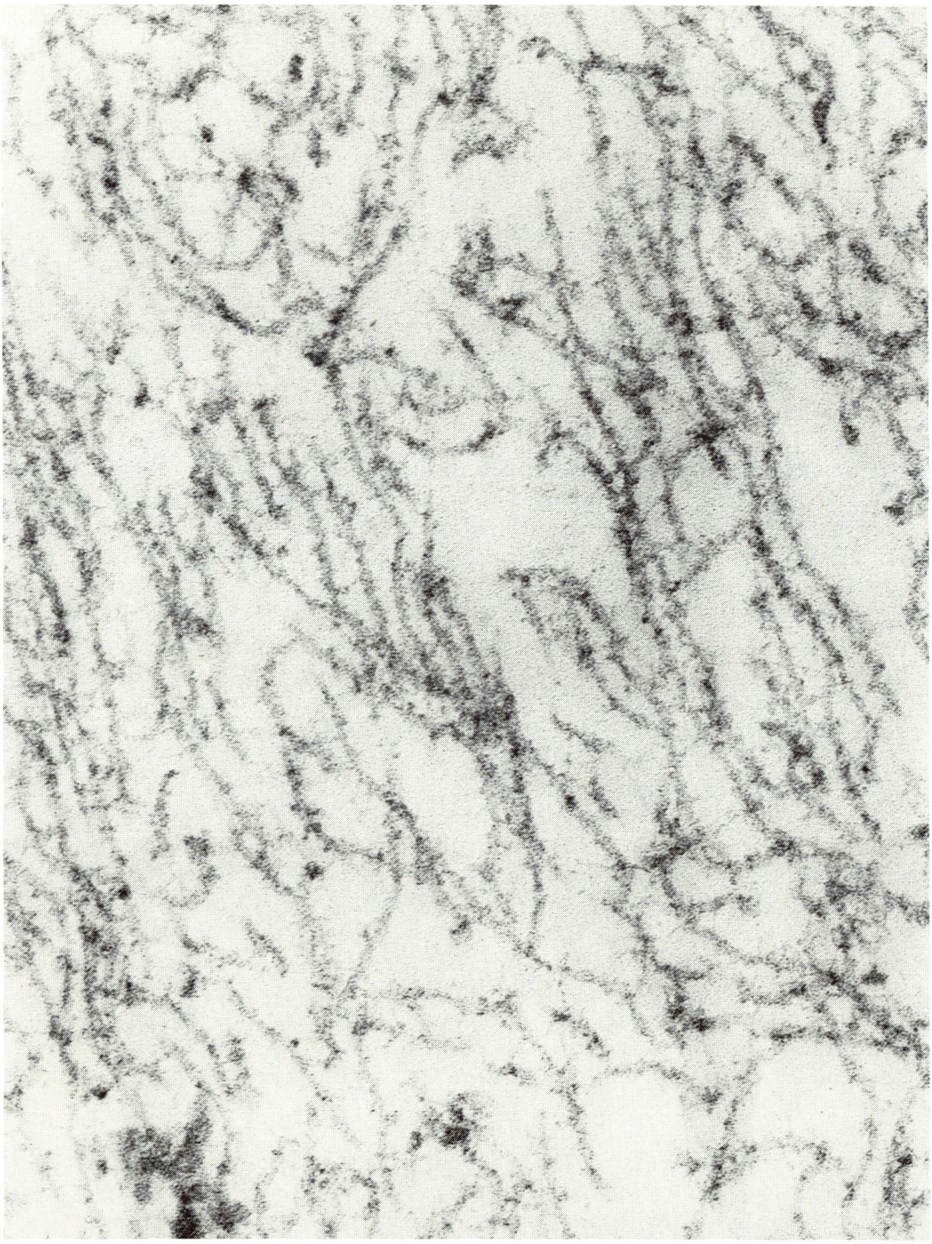

Fig. 3. Electron micrograph of purified, isolated axonal neurofilaments. Filaments are 90 Å in diameter.

but this does not diminish the general attractiveness of his model. Recently, Thoa et al. (30) have demonstrated that colchicine and vinblastine as well as cytochalasin, inhibit the release of norepinephrine and dopamine-β-hydroxylase in response to stimulation but have no effect on passive release. It is possible that a membrane associated tubulin-like protein might mediate such effects. Against this is the lack of significant particulate colchicine binding activity in adrenal, pancreatic islets or the thyroid. Nonetheless, the effects in nerve are more rapid and might depend on receptor localization and concentration.

Neurofilaments

A protein subunit has been purified from the axoplasm of the squid by Huneeus and Davison (13). The protein obtained had a molecular weight of approximately 70,000 and lacks colchicine and nucleotide binding. More recently we have developed a procedure for the isolation of neurofilaments from mammalian brain. One of the major obstacles to this purification was an initial difficulty in ascertaining the neuronal origin of the filaments. This difficulty was circumvented by first using the method of DeVries and Norton (4) to float myelinated axons free of other materials and then fractionating the axons to obtain neurofilaments. The filament obtained is insoluble in 0.6 M KCl and dissolves readily in 0.1 percent sodium dodecylsulfate and in 4 M urea (24). The subunit has a molecular weight of 60,000 and migrates with or slightly behind the more slowly moving of the two tubulin subunits. This protein differs considerably from the squid filament protein but specific similarities and differences have not been explored. No colchicine or nucleotide binding was demonstrable in the isolated filaments. These filaments (Fig. 3) have a diameter of 90 - 100 Å and should not be confused with 50 Å microfilaments which may be related to the muscle protein actin.

Summary

In this talk we have attempted to present a number of new and sometimes controversial observations from our laboratory in the hope that they may serve as an impetus to further work on the chemistry and function of tubules and filaments in brain. It seems clear that the bulk of this work still lies ahead.

ACKNOWLEDGEMENTS

This investigation has been supported by grant ##NS-08180 and Teacher-Investigator Award (MLS) ## 11,011 from the National

Institutes of Health.

The authors present addresses are :

- M.L.Shelanski and M.P.Daniels : Laboratory of Biochemical Genetics, National Heart and Lung Institute, National Institutes of Health, Bethesda, Md. 20014.
- H. Feit : Division of Neurology, Univ. of Colorado School of Medicine, Denver, Colo.
- R.W. Berry : Division of Biology, California Institute of Technology, Pasadena, Calif.

REFERENCES

1 BORISY, G.G. and TAYLOR, E.W., J. Cell Biol., 34 (1967) 525.
2 BORISY, G.G. and TAYLOR, E.W., J. Cell Biol., 34 (1967) 535.
3 BRYAN, J. and WILSON, L., Proc. Natl. Acad. Sci. US, 8 (1971) 1762.
4 DE VRIES, G.H. and NORTON, W.T., Science, in press.
5 FALXA, M.L. and GILL, T.J., Arch. Biochem. Biophys., 135 (1969) 194.
6 FEIT, H., Ph.D. Thesis, Dept. Mol. Biol., A. Einstein College of Medicine, 1971.
7 FEIT, H. and BARONDES, S.H., J. Neurochem., 17 (1970) 1355.
8 FEIT, H., DUTTON, G., BARONDES, S.H. and SHELANSKI, M.L., J. Cell Biol., 51 (1971) 138.
9 FEIT, H., SLUSAREK, L. and SHELANSKI, M.L., Proc. Natl. Acad. Sci. US, 68 (1971) 2028.
10 FERNANDEZ, H.L., BURTON, P.R. and SAMSON, F.E., J. Cell Biol., 51 (1971) 171.
11 GOODMAN, D.B.P., RASMUSSEN, H., DI BELLA, F. and GUTHROW, C.E., Jr., Proc. Natl. Acad. Sci. US, 67 (1970) 652.
12 GOTTSCHALK, A. in Glycoproteins, Their Composition, Structure and Function, Elsevier Publishing Co., Amsterdam, 1966.
13 HUNEEUS, F.C. and DAVISON, P.F., J. Mol. Biol., 52 (1970) 415.
14 JALFORS, U. and SMITH, D.S., Nature (Lond.), 224 (1969) 710.
15 LACEY, P.E., HOWELL, S.L., YOUNG, D.A. and FICK, C.J., Nature (Lond.) 219 (1968) 1177.
16 LAGNADO, J.R., LYONS, C. and WICKREMASINGHE, G., FEBS Letters, 3 (1971) 254.
17 MARANTZ, R., VENTILLA, M. and SHELANSKI, M.L., Science, 165 (1969) 498.
18 OLMSTED, J.B., CARLSON, K., KLEBE, R., RUDDLE, F. and ROSENBAUM, J., Proc. Natl. Acad. Sci. US, 65 (1970) 129.
19 OLMSTED, J.B., WITMAN, G.B., CARLSON, K. and ROSENBAUM, J.L. Proc. Natl. Acad. Sci. US, 68 (1971) 2273.
20 POISNER, A.M. and BERNSTEIN, J., J. Pharm. Exptl. Therap., 177 (1971) 102.
21 RAINE, C.S., GHETTI, B. and SHELANSKI, M.L., Brain Res., in press.

22 RASMUSSEN, H., Science, 170 (1970) 404.
23 RENAUD, F.L., ROWE, A.J. and GIBBONS, I.R., J. Cell Biol., 36 (1968) 79.
24 SHELANSKI, M.L., ALBERT, S., DE VRIES, G.H. and NORTON, W.T. Science, 174 (1971) 1242.
25 SHELANSKI, M.L. and FEIT, H., in Filaments and Tubules in the Nervous System in Structure and Function of Nervous Tissue (Ed. BOURNE, G.) Academic Press, New York, vol. VI, in press.
26 SHELANSKI, M.L. and TAYLOR, E.W., J. Cell Biol., 34 (1967) 549.
27 SHELANSKI, M.L. and TAYLOR, E.W., J. Cell Biol., 38 (1968) 304.
28 SJÖSTRAND, J. and KARLSSON, J.O., J. Neurochem., 18 (1971) 975.
29 STEPHENS, R.E., J. Mol. Biol., 47 (1970) 353.
30 THOA, N.B., WOOTEN. G.F., AXELROD, J. and KOPIN, I.J., Proc. Natl. Acad. Sci. US, in press.
31 WEISENBERG, R.C., BORISY, G.G. and TAYLOR, E.W., Biochemistry, 7 (1968) 1968.
32 WEISENBERG, R.C. and TIMASHEFF, S.N., Biochemistry, 9 (1970) 4110.
33 WILLIAMS, J.A. and WOLFF, J., Proc. Natl. Acad. Sci. US, 67 (1970) 1901.
34 WILSON, L., Biochemistry, 9 (1970) 4999.

MEMBRANE PROTEINS OF CATECHOLAMINE-STORING VESICLES IN ADRENAL MEDULLA AND SYMPATHETIC NERVES

H. WINKLER[+], Heide HÖRTNAGL[+], H. ASAMER[++] and H.PLATTNER[+++]
University of Innsbruck, [+]Department of Pharmacology, [++]Department of Medicine, [+++]Department of Electron Microscopy - 6020 Innsbruck (Austria)

The hormones of the adrenal medulla are stored in a specialized cell particle, the chromaffin granule. Since these organelles can be easily isolated, their biochemical characterisation is already well advanced (see Ref. 26). Less is known about the biochemical properties of the catecholamine-storing vesicles of sympathetic nerve (see Ref. 21). It has, however, already been shown that the vesicles of splenic nerve have two proteins in common with the medullary granules. One of these proteins, which has been called chromogranin A (5, 25), was first isolated from the soluble lysate of chromaffin granules (10, 29, 30). Its presence in the vesicles of bovine splenic nerve has been shown by immunological methods (3, 6, 9, 15). A second protein, common to both storage organelles, is the enzyme dopamine β-hydroxylase (6, 16, 21). It would not be surprising if the synaptic vesicles isolated from brain also have properties similar to those of the catecholamine storing vesicles of the peripheral tissues. Thus, results obtained on these catecholamine storing organelles might help to elucidate the biochemical and functional properties of the complex mixture of isolated synaptic vesicles which store several transmitters, only one of them being noradrenaline.

The present report deals mainly with the membrane proteins of the chromaffin granules in the adrenal medulla. These proteins have recently been named chromomembrins (34) in order to distinguish them from the soluble granule proteins, the so-called chromogranins. Some results on the membrane proteins of catecholamine storing vesicles of sympathetic nerve are also reported.

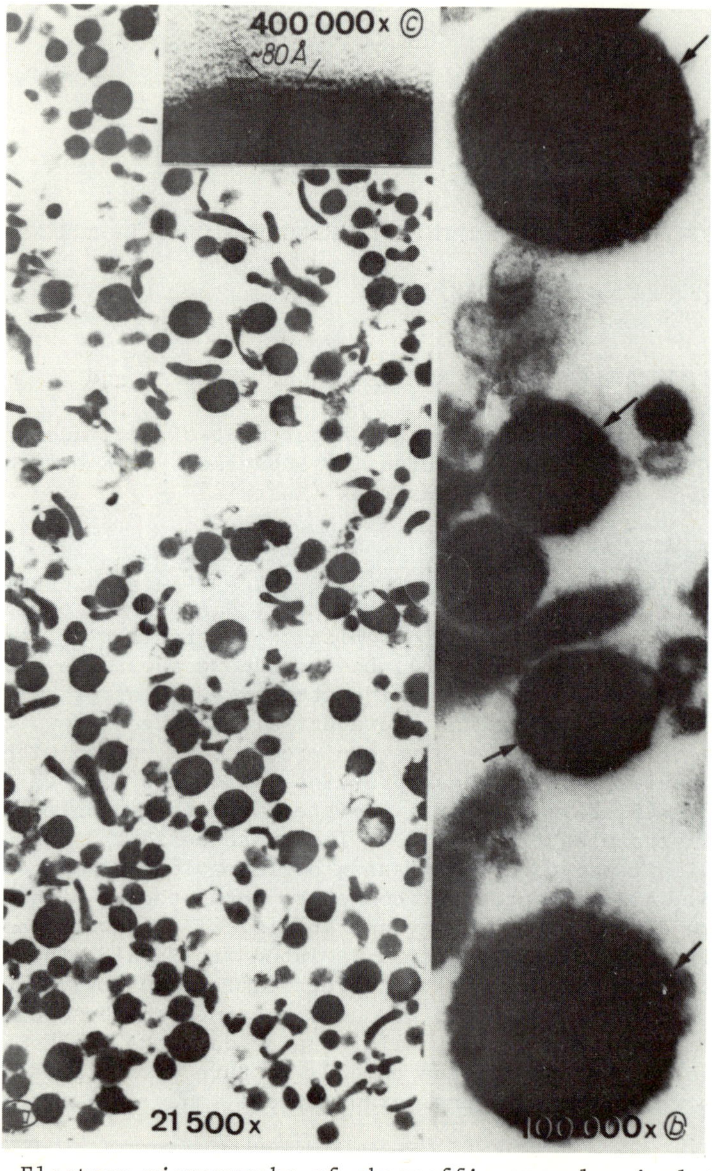

Fig. 1. Electron micrographs of chromaffin granules isolated according to Smith and Winkler (29). Fixation : glutaraldehyde followed by osmium tetroxide ; alkaline lead citrate stain. (a) Survey picture demonstrating the purity of the fraction. (b) Chromaffin granules at higher magnification. Areas, where the unit membranes show up in cross section, are indicated by arrows. (c) High resolution electron micrograph of the granule membrane demonstrating the unit membrane structure.

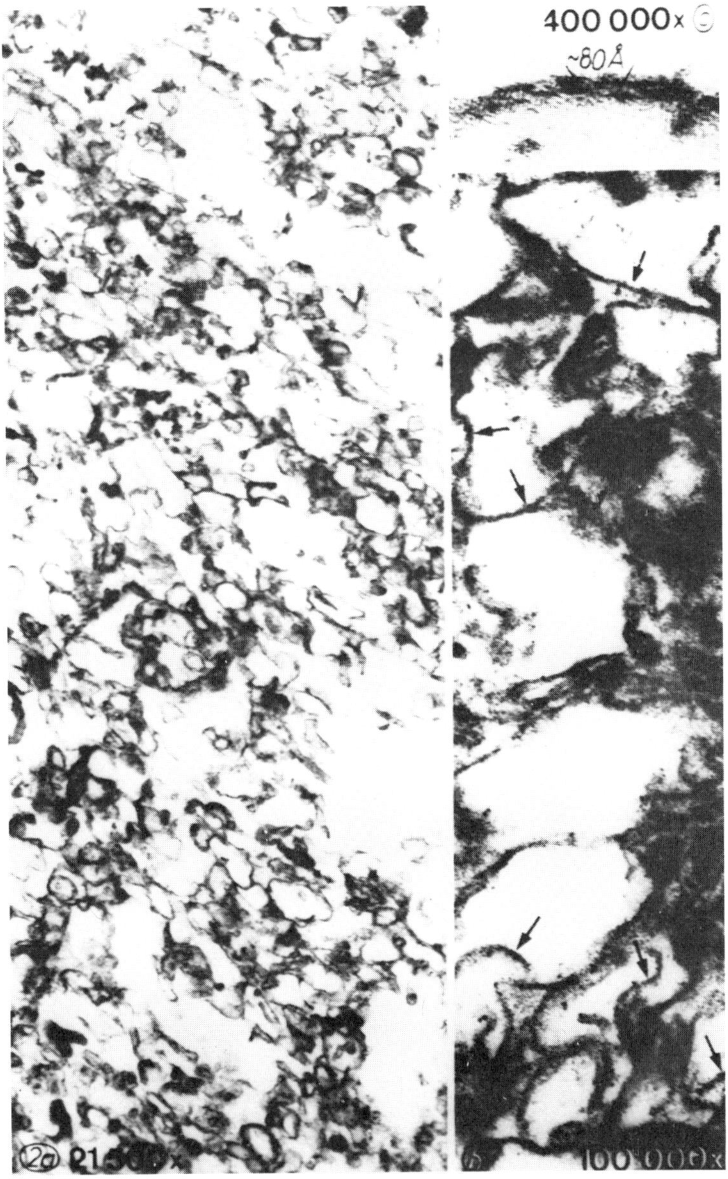

Fig. 2. Electron micrographs of membranes from chromaffin granules isolated according to Winkler et al. (35). Fixation and staining as in Fig. 1. (a) Survey picture demonstrating the absence of intact granules. (b) Higher magnification of the membrane fractions. Arrows point to regions where the unit membrane is cross sectionned. (c) High resolution electron micrograph of the isolated membranes.

The Isolation and Composition of Membranes from Chromaffin Granules

When isolated chromaffin granules (see Fig.1.) are resuspended in hypotonic buffer, the granules are lysed. Centrifugation of the suspension yields a sediment which consists mainly of the granule membranes. The catecholamines, the ATP and about 75 percent of the proteins, the chromogranins, remain in the supernatant (14). These constituents apparently make up the content of the granules,which can be seen in electron micrographs (Fig.1). The morphological appearance of the membrane pellet is characterized by the presence of bits and pieces of membranes. Intact granules could not be observed (Fig. 2).

The composition of the granule membranes which have been purified by several washing steps is given in Table 1. Main constituents are lipids and proteins. Only small amounts (see Table I) of chromogranin A contaminate the membrane preparation when purified as described by Winkler et al. (35). Three enzymes are found in the granule membranes, a Mg^{2+} activated ATPase (2,13,19), dopamine-β-hydroxylase (18,20,24) and a cytochrome (2,17) which has been classified either as a cytochrome b-559 (2,17) or as a cytochrome b-561 (8).

TABLE I

Composition of Membranes of Chromaffin Granules

Protein	1	mg
Lipid P	2.4	μmol
Cholesterol	1.66	μmol
Mg^{2+}-ATPase	1.8	μmol/h
Dopamine-β-hydroxylase	0.012	μmol/h
Cytochrome b-559	0.46	(E_{425})
Chromogranin A	0.04	mg

The data of this table are taken from Winkler et al. (35).

The possibility of a clear-cut separation of the membranes (containing the lipids) from the content of the granules has recently been questioned by Mylroie and König (23). These authors found that under their experimental conditions up to 80 percent of the lipids could be recovered as "soluble lipoproteins", the protein moiety of which was considered to be made up of the chromogranins. It was suggested that the freeze-thawing procedure used by Winkler et al. (35) in the purification of the membranes led to a delipidation of the chromogranins with the consequence that the lipids were rendered insoluble. However, Mylroie and König (23) routinely used Triton X-100 and ultrasonication in their procedure both of which are known to solubilize membranes. Some experiments were performed without ultrasonication or without the use of Triton but it is not clear from their paper whether both procedures were omitted together. Furthermore, their definition of solubility is uncertain since it is only stated that the granule suspensions were centrifuged for 30 min, but not with what centrifugal force. We therefore reinvestigated the isolation procedure for granule membranes. A sediment of isolated (29) chromaffin granules (corresponding to 3 g original tissue) was resuspended in 5 ml 0.001 M EDTA (pH 7.2; see Ref.23) for one hour in ice. Then the suspension was centrifuged for 180,000 g for 30 or 60 min. With both centrifugation times 76 percent of the protein were found in the supernatant. The "solubility" of the lipids varied. After 30 min centrifugation 13,3 percent of the lipids remained in the supernatant, after 60 min only 1.3 percent. Thus, the lipids are quantitatively sedimented if high centrifugal forces are used and if membrane solubilizing procedures are omitted. It seems likely that the very small amounts of lipids which are not sedimented are due to small, very light membrane bits.

Characterization of Membrane Proteins of Chromaffin Granules by Polyacrylamide Gel Electrophoresis

A resolution of the membrane protein has been achieved with polyacrylamide gel electrophoresis (16, 35). Membranes of chromaffin granules were either solubilized in sodium dodecylsulphate or in phenol-acetic acid-urea and then subjected to disc electrophoresis. Figure 3 gives a schematic drawing of the electrophoretic behaviour of the solubilized membranes in two different buffer systems. Such a pattern can only be achieved when detergents or phenol/urea are present in the gels during electrophoresis. It is, therefore, not surprising that Helle (11) did not obtain a comparable pattern since no detergent was present in the gels although the membranes were solubilized in Triton X-100.

The two main proteins present in the membranes of bovine chromaffin granules (Fig. 3) have been named chromomembrin A and B. Similar or identical components could be demonstrated in the chromaffin granule membranes of pig, horse and man (16). Chromomembrin B which migrates fast in the alkaline buffer sustem (Fig. 3a) remains near the origin in the acid system (Fig. 3b). This behaviour might be taken as an indication that this component is present in the acid system in a more expanded form. However, even when the gel concentration is reduced, chromomembrin B stays near the origin whereas the other proteins migrate, as to be expected, faster (Fig. 3c). We suggest, therefore, that chromomembrin B after having

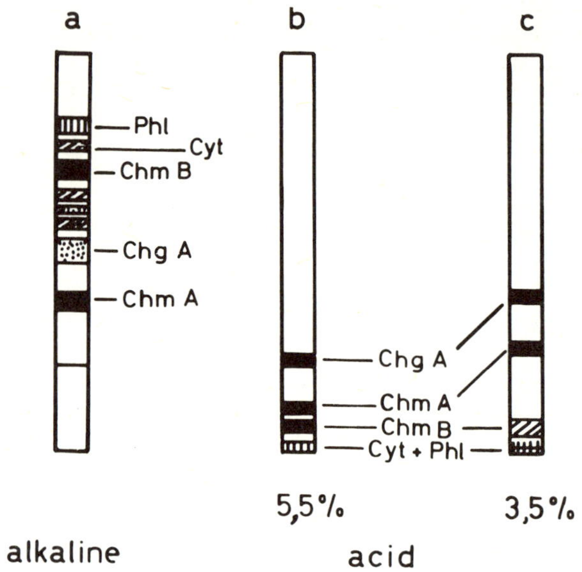

Fig. 3. Polyacrylamide disc electrophoresis of membranes of bovine granules (for methods see Ref. 35). (a) Membranes were dissolved in alkaline sodium dodecyl sulphate, and the gel contained sodium dodecylsulphate. The direction of migration was from the bottom to the top, which was the anode. (b) and (c) Membranes were dissolved in phenol -acetic acid- urea and applied to gels equilibrated with the solvent. The direction of migration was from the bottom to the top, which was the cathode. The gel concentration in (b) was 5.5 % in (c) 3.5 %. Chm. = chromomembrin ; Chg = chromogranin ; Cyt. = cytochrome ; Phl = phospholipids.

entered the gel becomes for unknown reasons insoluble and, therefore, stays near the origin, regardless of the pore size of the gel.

A third protein band, which stains relatively strongly, migrates to the same position as chromogranin A, the major protein of the soluble granule proteins. Immunological analysis (35) indicated that chromogranin A contributes only 4 % to the total proteins of the membrane preparation. It has been suggested that this protein stains relatively strongly due to its acidic nature, and, therefore, gives a very distinct band in the electrophoretic pattern. This small amount of chromogranin found in the membrane fraction is likely to represent some residual soluble material adhering to the membranes.

An electrophoretic analysis of the membranes of other organelles from bovine adrenal medulla was also performed (36). It could be shown that the electrophoretic pattern obtained from mitochondrial and microsomal membranes was completely different from that of the granule membranes. The conclusion was, therefore, put forward that chromomembrin A and B are specifically localized in chromaffin granules.

The Isolation of Membrane Proteins of Bovine Chromaffin Granules

A preparative separation of the membrane constituents could be achieved with Sephadex-chromatography in the presence of sodium dodecyl sulphate (16) (see Fig.4. for a typical elution pattern). Under these conditions the lipids were separated from the proteins and eluted as one peak. Chromomembrin A and chromomembrin B were well purified. Their amino acid composition resembled that of the total proteins of the isolated membranes but was completely different from that of the chromogranins (16).

It has been pointed out above that Winkler *et al.* (35) found only small amounts of chromogranin A associated with the membranes. Helle and Serck-Hansen(12), on the other hand, concluded from immunological experiments (double diffusion technique) that chromogranin A was a major component of the membranes, representing 75 percent of the proteins. However, the amino acid composition of the membrane proteins (13.9 % glutamic acid) reported by Helle and Serck-Hansen (12) already excluded such a high content of chromogranin A (26 % glutamic acid) in the membrane. Assuming that the other membrane proteins were devoid of glutamic acid, chromogranin A could make up 50 percent of the total proteins (12). However, the results reported above clearly show that the main membrane proteins contain about as much glutamic acid as the total membrane proteins. The membrane fraction can, therefore, contain only a small amount of chromogranin A, which is in agreement with the immunological studies (microcomplement fixation) of Winkler *et al.* (35).

Cytochrome b-559 could be identified in the eluates from the Sephadex columns by its specific absorption peak around 410 mμ (Ref.2). It is eluted (Fig. 4.) together with the phospholipids. A minor protein band has been demonstrated in these fractions by electrophoresis (16).

The enzyme ATPase and dopamine-β-hydroxylase were both inactivated by sodium dodecylsulphate. Therefore, different procedures had to be tried to characterize these enzymes. About 50 percent of the dopamine-β-hydroxylase activity present in chromaffin granules becomes soluble when the granules are lysed (4, 33, 35). It is not yet known whether this soluble enzyme and the membrane bound enzyme are identical. Recently, we discovered a rapid method to isolate this enzyme from the soluble lysate of chromaffin granules (Hörtnagl and Winkler, unpublished results). When N-cetylpyridiniumchloride

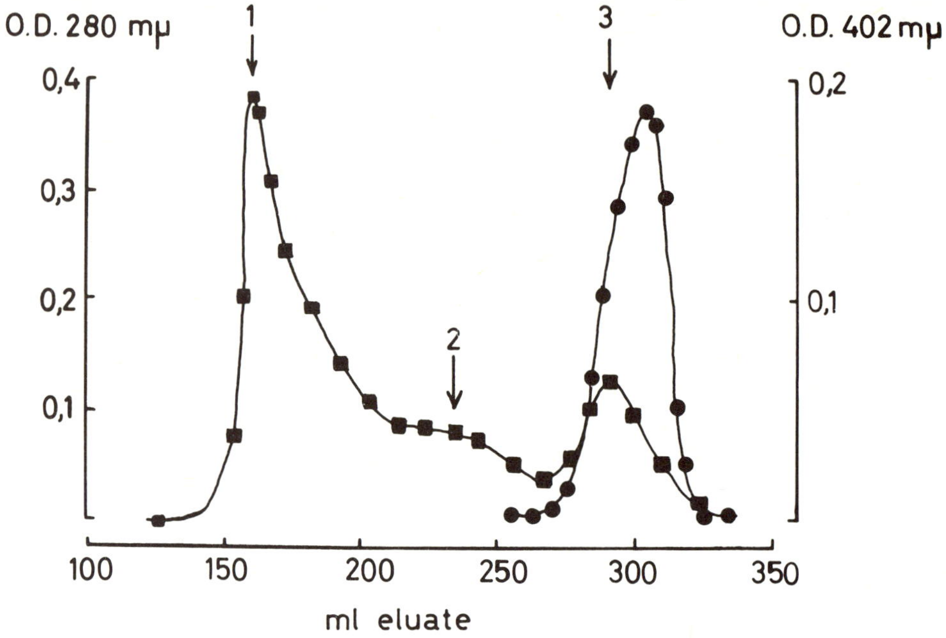

Fig. 4. Elution pattern of solubilized membranes (chromaffin granules) from Sephadex G-200 (16). The buffer used for elution contained sodium dodecylsulphate (1 %, w/v). Peak 1 contains mainly chromomembrin A, peak 2 chromomembrin B. The lipids are eluted in peak 3. The elution of cytochrome b-559 is indicated by the extinction readings at 402 m μ (uncorrected wavelength).
▪▪ extinction 280 mμ; ●●extinction 402 mμ.

(0.5 mg/mg protein) is added to the dialysed soluble proteins, an immediate precipitation of protein occurs. It was found that only a few percent of the total proteins, but about 70 percent of the dopamine-β-hydroxylase activity, remained in the supernatant. Electrophoresis of the supernatant revealed the presence of only that minor component of the chromogranins which migrates slower than chromogranin A (31) in disc electrophoresis. This component has been shown to be the enzyme dopamine-β-hydroxylase by direct analysis of portions of gels following electrophoresis (27) which is confirmed by the present results. Thus, this newly discovered precipitation method is a simple and quick procedure for isolating this enzyme. The amino acid composition of dopamine-β-hydroxylase is very similar to that of the membrane proteins but clearly different from that of the chromogranins. Thus, in the enzyme protein glutamic acid comprises 15 percent of the total amino acids, whereas in the chromogranins it comprises 26 percent. In polyacrylamide disc electrophoresis the purified enzyme behaves exactly like chromomembrin A. These results indicate the interesting possibility that the major protein component of the granule membranes is an enzyme, namely dopamine-β-hydroxylase.

Treatment of granule membranes with N-cetylpyridinium chloride (2.0 mg/mg protein) leads to solubilization of dopamine-β-hydroxylase. About 90 percent of the enzyme remained in the supernatant when the detergent-treated membranes were centrifuged for 60 min at 145,000 g. The solubilized membranes were fractionated on Sephadex G-200 columns in the presence of 0.01 percent N-cetylpyridinium-chloride. Chromomembrin A and dopamine-β-hydroxylase activity were eluted together. These studies are consistent with the suggestion that chromomembrin A possesses dopamine-β-hydroxylase activity.

Immunological Studies with Chromomembrin B

An antiserum was produced against chromomembrin B by injecting the purified protein into rabbits (1). The antiserum could not be tested by the double diffusion technique since it proved impossible to find a concentration of a detergent which solubilized the antigen without interfering with the antigen-antibody reaction. However, positive reactions were obtained with the indirect immunohistochemical method (1) and by microcomplement fixation (Winkler and Hörtnagl, unpublished work). Figure 5 gives an example of the immunohistological results. The medulla, in this case of horse adrenal, gave a positive reaction indicated by the fluorescence not observable in control sections. The cortex does not exhibit a specific fluorescence. Similar results were obtained in the adrenals of pig, rat and cat. Positive reactions were also obtained in sections of splenic nerve which contains mainly sympathetic fibres (1). Confirmation of these immunohistochemical results was sought by the use of microcomplement fixation (for method see Ref. 25). When isolated membranes of bovine chromaffin granules were tested, a fixation of complement could be demonstrated. The lowest amount of total membrane protein giving a fixation was about 300 ng. Addition of desoxycholate in various concentrations (up to 4 mg per mg of membrane protein) did not alter the results. Apparently the antigenic groups of chromomembrin B are present on the surface of the granule membranes and cannot be further exposed by solubilization with a detergent. No complement fixation occured when the soluble proteins of chromaffin granules were tested in concentrations up to 50 times higher then that used for the membrane proteins. Obviously there is not antigenic similarity between chromomembrin B and the chromogranins.

Complement fixation is at least a semi-quantitative method. It was, therefore, possible to measure the relative amounts of

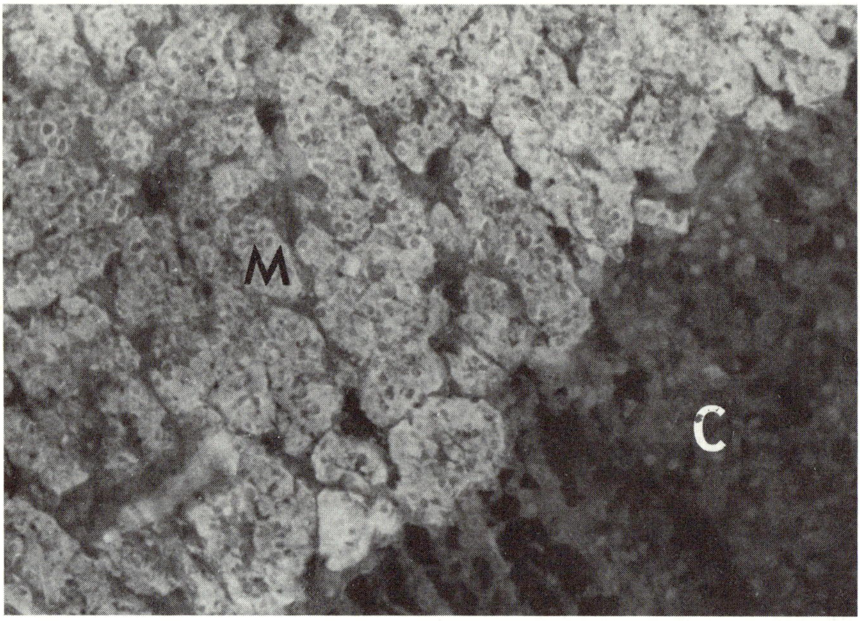

Fig. 5. Horse adrenal stained immunohistochemically. Cryostat sections (4 μ) were fixed in methanol, incubated with the antisera against component B and finally treated with fluorescein isothiocyanate-labelled goat anti-rabbit serum (Fa. Hyland). The medulla (M) fluoresces, but the cortex (C) does not. X 350.

TABLE II

Chromomembrin B in the Membranes of Subcellular Fractions of Bovine Adrenal Medulla

	Chr. B	DBH	Protein
Chrom. Gran.	100	100	100
Mitochondria	10.5	6	100
Microsomes	17.5	17	100

Subcellular fractions of bovine adrenal medulla were obtained as described by Smith and Winkler (28). The membranes of these fractions were isolated by the method of Winkler <u>et al</u>. (35). Chromomembrin B (Chr.B) was measured by micro-complement fixation, protein by the Biuret method and dopamine-β-hydroxylase (DBH) as described previously (16). The values for chromaffin granules and for protein (membrane) were taken arbitrarily as 100.

chromomembrin B in the various cell fractions of bovine adrenal medulla. Table II demonstrates that the antigen is specifically localized in the chromaffin granules. In the mitochondrial and microsomal fraction it is present in amounts comparable to that of dopamine-β-hydroxylase. It is known that both these fractions are slightly contaminated with both intact chromaffin granules and with membranes of damaged granules. Thus, the immunological results confirm the biochemical studies, reported above, demonstrating that chromomembrin B is not present in the mitochondrial and microsomal membranes.

The subcellular fractions of bovine splenic nerve were also tested for the presence of this chromomembrin. It has been shown previously that the noradrenaline vesicles can be separated from microsomal elements by density gradient centrifugation (16). The distribution of dopamine-β-hydroxylase and that of chromomembrin B was parallel in the various fractions of the gradient. This confirms the immunohistological results, but shows in addition that chromomembrin B is localized in the catecholamine-storing vesicles. It seems, therefore, that the membrane of the nerve vesicles may be quite similar to that of the adrenal medullary granules. It will be interesting to see whether other hormone and transmitter storing organelles possess also this membrane protein. Preliminary results indicate that it is present in the posterior hypophysis, probably in the hormone storing particles (Hörtnagl and Winkler, unpublished work).

Localization of the Proteins within the Membrane

A knowledge of the sub-membranous localization of these proteins is likely to provide some insight into their functional importance. Unfortunately, only a few results which appear relevant to this section are available.

A suspension of membranes of chromaffin granules was incubated together with trypsin or pronase (Hörtnagl, Winkler and Smith, unpublished results). The membranes were then sedimented with high speed centrifugation. Incubation with trypsin (up to 0.5 mg/mg protein) resulted in the disappearance of up to 40 percent of the membrane proteins, whereas with pronase (0.5 mg/mg protein) up to 50 percent could be digested. Disc electrophoresis of the membrane proteins after protease treatment revealed that chromomembrin A and B were still present. The band corresponding to chromogranin A had disappeared. A further breakdown of membrane proteins could not be achieved when intact membranes were used. However, when the membranes were first solubilized in sodium dodecyl sulphate, pronase digested all proteins as judged by disc electrophoresis. Pronase treatment of the membranes resulted in a complete loss of ATPase activity (Smith, Hörtnagl and Winkler, unpublished results) which might indicate that this enzyme is localized on the surface of the membranes. The phospholipids, on the other hand, can also be attacked by enzymes, since phospholipase C treatment resulted in the breakdown of 75 percent of the phospholipids. Similar results have already been reported for the membranes of erythrocytes (22) and microsomes (32).

Conclusions

The membrane proteins of chromaffin granules have been characterized. Two main components (chromomembrin A and B) have been isolated. The presence of chromomembrin B in the catecholamine-storing vesicles of sympathetic nerve could be demonstrated by immunological techniques. It will be interesting to see whether the membranes of other hormone and transmitter storing organelles have similar membrane proteins.

ACKNOWLEDGEMENTS

This work reported in this paper was made possible by the support of the Fonds zur Förderung der wissenschaftlichen Forschung (Austria).

REFERENCES

1 ASAMER, H., HÖRTNAGL, H. and WINKLER, H., Arch. Pharmak., 270 (1971) 87.
2 BANKS, P., Biochem. J., 95 (1965) 490.
3 BANKS, P., HELLE, K.B. and MAYOR, D., Mol. Pharmacol., 5 (1969) 210.
4 BELPAIRE, F. and LADURON, P., Biochem. Pharmacol., 17 (1968) 411.
5 BLASCHKO, H., COMLINE, R.S., SCHNEIDER, F.H., SILVER, M. and SMITH, A.D., Nature (Lond.), 215 (1967) 58.
6 DE POTTER, W.P., SMITH, A.D. and DE SCHAEPDRYVER, A.F., Tissue Cell, 2 (1970) 529.
7 DUCH, D.S., VIVEROS, O.H. and KIRSHNER, N., Biochem. Pharmacol., 17 (1968) 255.
8 FLATMARK, T., TERLAND, O. and HELLE, K.B., Biochim. Biophys. Acta, 226 (1971) 9.
9 GEFFEN, L.B., LIVETT, B.G. and RUSH, R.A., J. Physiol. (Lond.), 204 (1969) 593.
10 HELLE, K.B., Mol. Pharmacol., 2 (1966) 298.
11 HELLE, K.B., Biochim. Biophys. Acta, 245 (1971) 80.
12 HELLE, K.B. and SERCK-HANSSEN, G., Pharmacol. Res. Commun., 1 (1969) 25.
13 HILLARP, N.A., Acta Physiol. Scand., 42 (1958) 144.

14 HILLARP, N.A., Acta Physiol. Scand., 43 (1958) 82.
15 HOPWOOD, D., Histochemie, 13 (1968) 323.
16 HÖRTNAGL, H., WINKLER, H., SCHÖPF, J.A.L. and HOHENWALLNER, W. Biochem. J., 122 (1971) 299.
17 ICHIKAWA, Y. and YAMANO, T., Acta Physiol. Scand., 45 (1959) 328.
18 KIRSHNER, N., Pharmacol. Rev., 11 (1959) 350.
19 KIRSHNER, N., KIRSHNER, A.G. and KAMIN, D.L., Biochim. Biophys. Acta, 113 (1966) 332.
20 LADURON, P. and BELPAIRE, F., Biochem. Pharmacol., 17 (1968) 1127.
21 LAGERCRANTZ, H., Acta Physiol. Scand. Suppl. 366 (1971).
22 LENARD, J. and SINGER S.J., Science, 195 (1968) 738.
23 MYLROIE, R. and KÖNIG, H., Febs Letters, 12 (1971) 121.
24 OKA, M., KAJIKAWA, K., OHUCHI, T., YOSHIDA, H. and IMAIZUMI, R. Life Sci., 6 (1967) 461.
25 SCHNEIDER, F.H., SMITH, A.D. and WINKLER, H., Brit. J. Pharmacol. Chemotherap., 31 (1967) 94.
26 SMITH, A.D., in The Interaction of Drugs and Subcellular Components on Animal Cells (Ed. CAMPELL, P.N.) Churchill, London, 1968, p.239.
27 SMITH, A.D., DE POTTER, W.P., MOERMAN, E.J. and SCHAEPDRYVER, A.F., Tissue Cell, 2 (1970) 547.
28 SMITH, A.D. and WINKLER, H., J. Physiol. (Lond.) 183 (1966) 179.
29 SMITH, A.D. and WINKLER, H., Biochem. J., 103 (1967) 483.
30 SMITH, W.J. and KIRSHNER, N., Mol. Pharmacol., 3 (1967) 52.
31 STRIEDER, N., ZIEGLER, E., WINKLER, H. and SMITH, A.D., Biochem. Pharmacol., 17 (1968) 1553.
32 TRUMP, B.F., DUTTERA, S.M., BYRNE, W.L. and ARSTILA, A.U., Proc. Natl. Acad. Sci. US, 66 (1970) 433.
33 VIVEROS, O.H., ARQUEROS, L., CONNETT, R.J. and KIRSHNER, N., Mol. Pharmacol., 5 (1969) 69.
34 WINKLER, H., Phil. Trans. Roy. Soc. B 261 (1971) 293.
35 WINKLER, H., HÖRTNAGL, H., HÖRTNAGL, Heide and SMITH, A.D., Biochem. J., 118 (1970) 303.

BIOLOGICAL PROPERTIES OF THE NERVE GROWTH FACTOR

Pietro U. ANGELETTI

Istituto Superiore di Sanità e Laboratorio di Biologia Cellulare, Roma (Italy)

While during early developmental stages, both sensory and sympathetic nerve cells are receptive to the Nerve Growth Factor (NGF) only the sympathetic neurons retain this prerogative at later stages and in adult life. These differences between the two types of nerve cells, already apparent from earlier studies by Levi-Montalcini in chick embryo, become more evident when the NGF effect is studied in newborn and adult mice. Neither in newborn nor in adult mouse does NGF elicit an overgrowth of sensory ganglia ; the sympathetic ganglia, on the contrary, undergo a dramatic increase in size (3, 5). When highly purified NGF, the NGF molecule whose primary structure has been established, is injected in newborn mice or rats at a dose of 10 μg/g body weight, a progressive enlargement of all sympathetic ganglia ensues. The size usually reaches 4-5 fold the normal values after a week or ten days treatment (Fig. 1).Even more dramatic overgrowth can be obtained by using larger doses of NGF for longer periods of time (4). From histological examination the size increase of each ganglion appears to be due to both an increase in cell number and to an increase in cell size (Fig. 2). Therefore, mitotic activity, as determined between the third and ninth day after birth, is considerably increased in NGF-treated mice and rats.

Around 9 days, mitotic activity finishes in sympathetic neurons and in subsequent stages up to the adult life, the NGF only increases the size of individual neurons, but does not further increase their number. In adult animals, daily injections of NGF (10 μg/g) result in a two-three fold enlargment of the superior cervical ganglia as well as of all para- and pre-vertebral ganglia ; the nerve cells are markedly increased in size, with enlarged cytoplasm and very prominent nucleoli. Upon cessation of the treatment, the

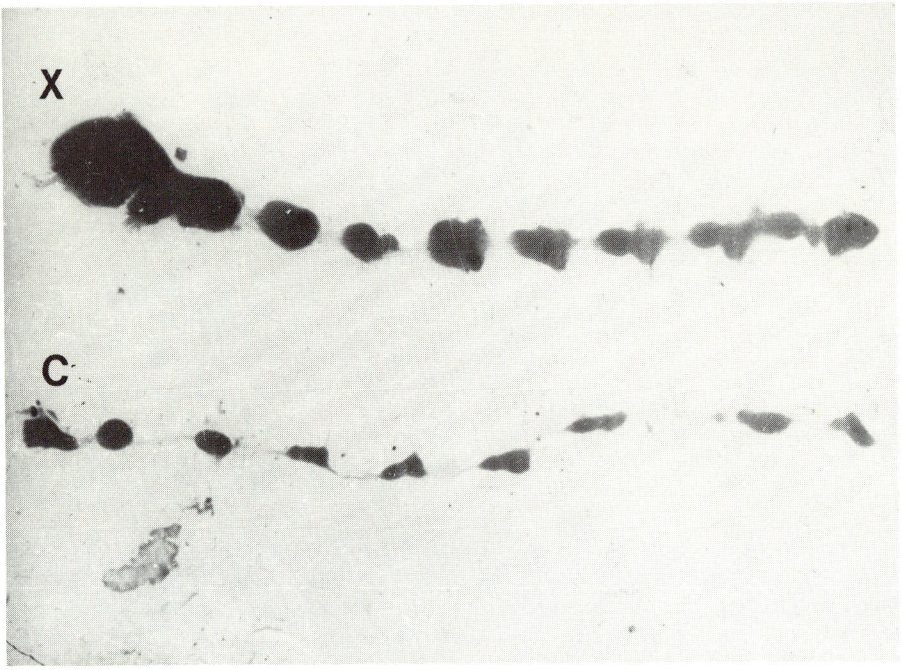

Fig. 1. Sympathetic chain-ganglia from control (C) and NGF-treated mice (X). Note the marked size increase of each ganglion.

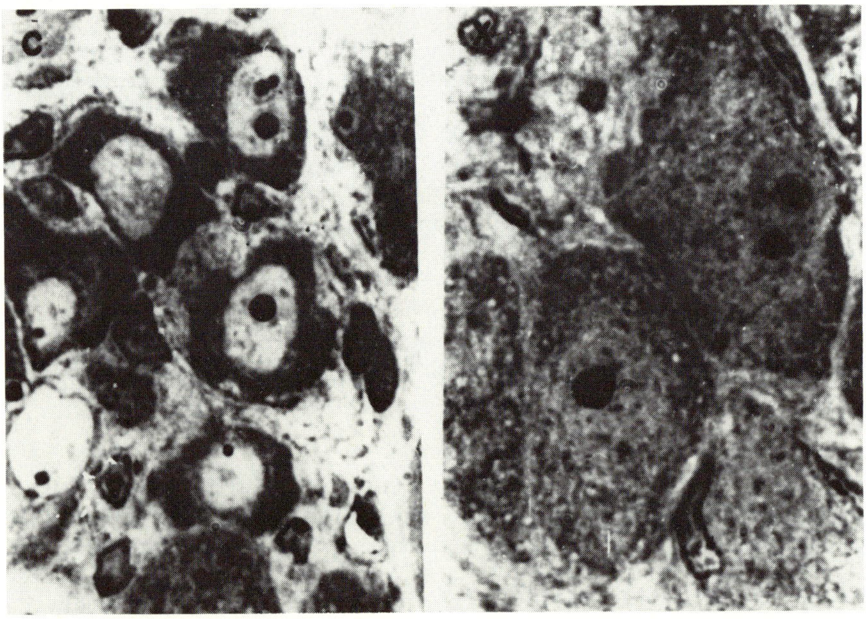

Fig. 2. Cross-sections of superior cervical ganglia from cervical ganglia from control (C) and NGF-treated mice (X). Note the size increase of neurons, while satellite cells are not affected.

cells gradually return to a normal size within a week period. The hypertrophic effect induced by the NGF appears therefore to be reversible.

Effect of NGF on Catecholamine Metabolism

Following NGF treatment a general increase of synthetic and oxidative processes is observed in sympathetic ganglia. Recent evidence has been given for a specific stimulation of enzymes involved in the metabolism of the sympathetic neurotransmitter (7). Newborn rats were injected for ten days with NGF (10 μg/g body weight). At the end of the treatment the superior cervical ganglia, from control and treated rats, were dissected out, weighed and homogenized. Total proteins, tyrosine hydroxylase, dopamine-β-hydroxylase, dopa-decarboxylase and monoamine oxidase, were measured in the extracts. The results obtained, in collaboration with Dr. H. Thoenen, can be summarized as follows : on a protein basis, the average protein content per pair of ganglia was 65 ± 4 μg in controls and 225 ± 10 μg in NGF treated animals.

The total activity of tyrosine hydroxylase (product formed/h/pair of ganglia) was increased 18-fold in NGF-treated animals and dopamine-β-hydroxylase activity was increased 13-fold. Dopa decarboxylase and monoamine oxydase activities were increased only to 4.5-fold and 2.5-fold respectively. The ratio between the specific activity of NGF-treated animals and controls was 5.4 for tyrosine hydroxylase, 3.6 for dopamine-β-hydroxylase, 1.5 for dopa decarboxylase and 1.2 for monoamine oxidase. The marked increase in the activity of tyrosine hydroxylase and dopamine-β-hydroxylase was shown not to be due to direct activation of the enzymes nor to the disappearance of some inhibitor, but rather to an induction of new enzyme molecules, stimulated by NGF.

These results are in line with previous findings by Crain *et al.*(2) that upon NGF-treatment there is a marked rise in the noradrenaline concentration in sympathetic ganglia.

It appears, therefore, that the NGF does not simply cause an overgrowth of the receptive nerve cells, but also specifically enhances the activity of those enzymes directly involved in the synthesis of noradrenaline.

Localization of NGF in Subcellular Fractions of Peripheral Tissues by Micro Complement Fixation

The hypothesis that the NGF may play a physiological role, related with the functional activity of the sympathetic cells, is indirectly supported by the recent experiments aimed to localize NGF in peripheral tissues. It has long been known that aside from the two richest sources, mouse submaxillary glands and snake venoms, NGF is also present in other tissues and body fluids (1, 4). The widespread presence of NGF in various tissue extracts and in blood has always been determined by the tissue culture method.

Recently we have used micro-complement fixation as a sensitive and highly specific method of assay of NGF in various peripheral tissues of mice and rats. Complement fixation was carried out by the micromethod of Wasserman and Levine (8). Tissue homogenates from mice (Swiss Strain) and rats (Wistar) were prepared in isotonic sucrose containing Mg^{++} (0.001 M) and sodium phosphate buffer (pH 7.4). The homogenates were subjected to differential centrifugation. The pellet recovered after centrifugation at 100,000 x g for 2 h was resuspended in sucrose and centrifuged again. After the second centrifugation the microsomal pellet was resuspended in 0.01 M phosphate buffer(pH 7.4) and dialysed overnight against the same buffer. Proteins were measured by the Lowry method using bovine serum albumin as standard. Fig. 3 shows a micro-complement fixation curve obtained with purified mouse NGF and its antiserum. As it can be seen in the first part of the curve (zone of antibody excess) the percent of complement fixation increases almost linearly with increasing the antigen concentration. The sensitivity of the method used in the present investigation allowed a reliable assay of NGF from 2 to 10 mμg ; the sensitivity could be further increased almost

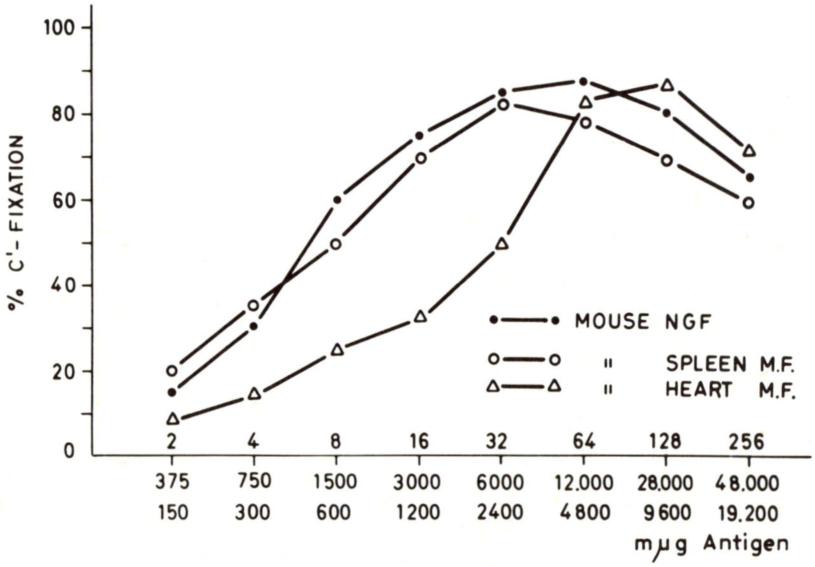

Fig. 3. Complement fixation curves with mouse NGF-antiserum and purified NGF, mouse heart microsomal fraction and mouse spleen microsomal fraction.

ten times by using the modification suggested by Moore and Perez (6). This however proved not to be necessary in the present experiments. Attempts to measure the NGF concentration in crude extracts of rat or mouse heart, spleen and kidney were not successful because of the extensive haemolytic and/or anticomplementary effects of the extracts. Similarly negative results were obtained when micro-complement fixation was carried out with the precipitate at 12,000 x g resuspended in phosphate buffer.

A completely different picture was observed when NGF was assayed in the microsomal pellets from different peripheral tissues. In all instances, with mouse tissues, clear cut complement fixation curves were obtained with a maximum of fixation reaching 85 to 95 percent, as occurs using the pure NGF antigen (Fig. 3). A semi-quantitative assay of the NGF concentration in the various microsomal preparations was then possible by calculating the amount of protein required to obtain 50 percent complement fixation, both with the microsomal fraction and with the pure NGF. The results are shown in table I. As it can be seen, NGF was present in the microsomal fraction of all mouse tissues examined in relatively

TABLE I

Relative Concentration of NGF in the Microsomal Fractions of Mouse Tissues, Calculated from the Amount of Protein Required to Give a 50 % Complement Fixation with Purified NGF and with the Microsomal Fraction

Tissue	N° Exp.	NGF mμg	Microsomal fraction mμg	NGF concentration mg %
Heart	1	6	2,400	0.25
	2	4.5	3,300	0.14
	3	7	1,400	0.50
	4	5.4	3,200	0.17
Spleen	1	7	3,500	0.20
	2	4.5	3,200	0.14
	3	5.4	1,800	0.30
Kidney	1	4.5	3,800	0.12
	2	5.4	2,200	0.24
	3	6.4	4,500	0.14

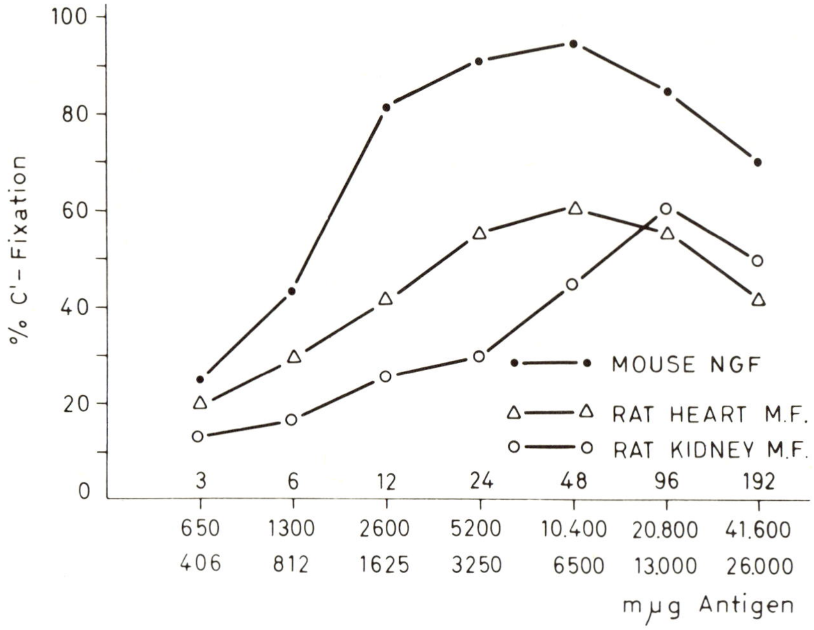

Fig. 4. Complement fixation curves with mouse NGF-antiserum and purified mouse NGF, rat-heart microsomal fraction and rat-kidney microsomal fraction.

high concentrations, ranging from 0.1 to 0.5 percent of the total protein. When the preparations were checked in tissue culture,NGF activity was also clearly observed. The activity displayed in tissue culture however, was from 10 to 100 times less than expected from the NGF concentration in the microsomal samples as measured by complement fixation. In rat tissues NGF was also detected in the microsomal fractions. Complement fixation curves were run with the antiserum to mouse NGF, and the maximum complement fixation never reached values higher than 60 to 65 percent. This vertical shift in the C-fixation curve would indicate a cross-reaction between two similar but not identical NGF molecules in mouse and rat tissues (Fig. 4).

Table II shows the relative concentration of NGF in rat kidney, heart and spleen microsomal fractions. The presence of NGF activity in these rat preparations was confirmed by the tissue culture method and again it was found that, at corresponding micro-complement fixation NGF concentrations, the activity of the fractions was about 100 times less than that observed with purified NGF from mouse submaxillary gland.

The results of these experiments confirm the widespread presence of NGF in several peripheral tissues. The protein appears to be

TABLE II

NGF Concentration in the Microsomal Fractions from Rat Tissues, Determined as in Table I

Tissue	N° Exp.	NGF mμg	Microsomal fraction mμg	NGF concentration mg %
Heart	1	7	4,600	0.15
	2	4.5	6,200	0.075
Kidney	1	7	9,500	0.08
	2	4.5	5,200	0.08
Spleen	1	7	6,400	0.1
	2	4.5	7,300	0.06

mostly present in a particulate form associated with the microsomal fraction. The apparent discrepancy between the concentration of NGF in this fraction as measured immunochemically, and that calculated from the tissue culture method, could be explained by several alternative hypotheses. In its membrane-bound form the NGF could be less active, or it could be associated with other regulatory proteins which limit its activity in the "*in vitro* test" , or its activity could be inhibited by other molecules present in the microsomal fraction.

Concluding Remarks

The main biological property of NGF *in vivo* is the selective growth stimulation of peripheral sympathetic system. The effect is most striking in newborn animals where hyperplastic and hypertrophic effects are obtained ; in adult animals only the hypertrophic effect is elicited. The growth effect is reversible and the nerve cells return to their normal size upon interruption of the treatment. The results reported above show that the NGF stimulation of the receptive nerve cells causes a selective induction of those enzymes which are rate-limiting in the synthesis of noradrenaline. Thus, NGF not only stimulates the growth of the receptive nerve cells but also markedly stimulates their functional activity. The finding that NGF is present in significant amounts in microsomal fractions of various peripheral tissues, suggests the possibility that under normal conditions it may play a role related to the function of the sympathetic neurons. Experiments along this line are now in progress.

REFERENCES

1 BUCKER, E.D., SCHENKEIN, I. and BONE, J.L., Cancer Res., 20 (1960) 1220.

2 CRAIN, S., BENITEZ, H. and VATTER, A.E., Ann. N.Y. Acad. Sci., 118 (1964) 206.

3 LEVI-MONTALCINI, R., Harvey Lecture Sci., 60 (1966) 217.

4 LEVI-MONTALCINI, R. and ANGELETTI, P.U., Physiol. Rev., 48 (1968) 534.

5 LEVI-MONTALCINI, R. and BOOKER, B., Proc. Natl.Acad. Sci. US, 46 (1960) 302.

6 MOORE, B.W. and PEREZ, V.J., J. Immun., 96 (1966) 1000.

7 THOENEN, H., ANGELETTI, P.U., LEVI-MONTALCINI, R. and KETTLER, R., Proc. Natl. Acad. Sci. US, 68 (1971) 1598.

8 WASSERMAN, E. and LEVINE, L., J. Immun., 87 (1961) 290.

THE PHYSICAL AND BIOLOGICAL PROPERTIES OF 7S AND β-NGF FROM THE MOUSE SUBMAXILLARY GLAND

J.R. PEREZ-POLO, W.W.W. DE JONG, Daniel STRAUS and
E.M. SHOOTER
Departments of Genetics and Biochemistry and the Lt.
Joseph P.Kennedy Laboratories for Molecular Medicine,
Stanford University School of Medicine,Stanford, Calif.

Recent progress with the Nerve Growth Factor (NGF) from the mouse submaxillary gland has concentrated on the further characterization of the two major molecular weight forms in which it exists, namely the 7S NGF complex present in homogenates of the gland and the smaller basic proteins which may be obtained from the complex by dissociation. Of the latter, two species, the β subunit isolated from 7S NGF (21) and 2.5 S NGF, obtained from the complex at an intermediate stage in its purification (4) have been investigated in detail. Such studies are the basis for an examination of structure-function relationships and for a variety of approaches to determine the mechanism of action of NGF. Although the morphological effects of NGF are well described, the sequence of biochemical events underlying these phenomena remains obscure.

In the absence of added NGF, explanted chick embryonic sensory and sympathetic ganglia,or disssociated nerve cells from these ganglia,put out very few nerve fibers and degenerate over a 24 hour period (10). With added NGF at appropriate concentrations, the ganglia produce an uniform fibrillar halo and the dissociated nerve cells survive and produce large numbers of nerve fibers (10, 11). The incorporation of labelled amino acids or of uridine into the cultured ganglia shows an abrupt decrease after 6 to 12 h incubation when NGF is absent, but not when it is present, giving rise to significant increases in incorporation in NGF containing cultures over periods of 22 h (1, 13). Smaller increases in incorporation are noted during the first few hours of incubation. Whether these reflect primary events of NGF interaction with the responsive neurons or are a consequence of cellular deterioration in the absence of the growth factor cannot yet be determined (13). Since nearly complete inhibition of RNA synthesis in sympathetic ganglia does not inhibit fiber outgrowth, it follows that the NGF induced increase in RNA synthesis is not by itself the cause of fiber growth (13). In a recent study, Thoenen et al. (18) have, however, shown that NGF does affect certain metabolic pathways in adrenergic neurons by selectively inducing tyrosine hydroxylase and dopamine-β- hydroxylase.

It has been known for some time that insulin in high concentrations can stimulate energy metabolism and anabolic processes in sensory and sympathetic ganglia leading to their preservation in cultures (1, 12). Insulin does not, however, produce fiber outgrowth. Two other agents which can, like NGF, induce fiber outgrowth have recently been described. Roisen et al. (15) have shown that cyclic AMP or its dibutyryl derivative increases both the number and length of axons from cultured embryonic chick dorsal root ganglia.The other substance is concanavalin A, which in relatively high concentrations compared to NGF, stimulates fiber outgrowth from chick embryonic ganglia (19). The effect of concanavalin A is inhibited by α-methylmannoside. The further evaluation of the effects of these two substances compared to those of NGF will be of considerable value.

The Physical and Biological Properties of 7S NGF

The 7S NGF species constitute a family of closely related protein complexes which may be isolated from extracts of the adult male mouse submaxillary gland at neutral pH (16,20). Their properties have been recently reviewed in detail (14,22) and recent work has been concerned with attempts to understand their physiological significance. In two growth factors isolated from the submaxillary gland, 7S NGF and the epidermal growth factor (EGF), the protein which elicits specific growth effects is associated with an arginine esteropeptidase (5,17). In the case of EGF, the EGF protein is relatively small (6,400 molecular weight) and acidic, while in 7S NGF, the NGF protein, or β-NGF subunit, is somewhat larger and basic. 7S NGF also contains a third type of subunit, the α subunit. Several arginine esteropeptidases are also found in the submaxillary gland which are not associated with other proteins, but which do appear to be important in terms of the growth and differentiation of certain cell types (3,8,9).In spite of their differences in physical and biological properties, the EGF and β-NGF proteins have one common characteristic,namely that the single polypeptide chain of EGF and the two identical chains in β-NGF have C-terminal arginine residues. Taylor et al.(17) were the first to suggest on the basis of this data that the EGF protein may be derived from a larger precursor by the selective action of the arginine esteropeptidase.A similar conclusion can be drawn for β-NGF (see also Ref.2) with one important distinction. The C-terminal arginine residue in EGF is the only arginine residue in the protein while this is not so for β-NGF.The EGF complex, consisting of the EGF and the arginine esteropeptidase subunits, retains the full enzymatic activity of the latter while on the other hand, the 7S NGF complex has very low or no arginine esteropeptidase activity. It could therefore be argued that a function of the third α subunit of 7S NGF is to provide an environment in the 7S NGF complex for completely inhibiting the esteropeptidase activity and thus preventing further cleavage of the β-NGF at internal arginine residues. Such a mechanism is not required in the EGF complex. Aside from these speculations about the role of the arginine esteropeptidase, it should be noted that it has one other property. Greene, Tomita and Varon (7) have shown by cell counts and incorporation of radiolabelled precursors into DNA that the esteropeptidase subunit releases chick embryo body wall fibroblasts from contact inhibition and have discussed the implications of this finding in terms of the regulation of cell growth.

The considerations noted above lead to the further conclusion that the binding of the arginine esteropeptidase or γ subunit with its two neighboring subunits, the putative product of its reaction, the β subunit, and the α subunit should be greater than that of a typical enzyme substrate complex and this appears to be so. On the one hand, differences in NGF activity between 7S NGF and the β-NGF subunit are detected in the usual bioassay procedure (11) where the protein concentration is of the order of 10^{-10} M (21), suggesting a remarkable stability for the complex, while on the other, the subunits display a high degree of affinity for one another even at low concentration in the presence of non-NGF proteins (14). The experimental procedure for demonstrating this last point involved isolation of 7S NGF from extracts of submaxillary gland pretreated by exposure to alkaline pH to dissociate the 7S NGF complex and subsequent redialysis to neutral pH to allow for its reformation, if possible under these conditions, from the subunits. The recovery of 7S NGF from such extracts was, in fact, slightly in excess of that from submaxillary gland extracts stood for the same length of time at neutral pH, both recoveries being about 60 % of normal. It is reasonable to assume that the lower than normal recoveries in each instance results from proteolytic activity in the gland extract.The similarity of the physical properties and subunit composition of the reformed as compared to native 7S NGF also testifies to the high subunit affinities.A new method for isolating 7S NGF has been devised based on this property (14).

The Biological Properties of β-NGF

Of the three subunits in 7S NGF, only the β subunit or β-NGF elicits fiber outgrowth from embryonic chick sensory and sympathetic ganglia. Because approximately the same concentrations,10-20 ng/ml, of β-NGF and 7S NGF give optimal response in the bioassay,it would appear that the interaction of the γ and α subunits with β-NGF either enhances or stabilizes its specific NGF activity. Further attempts to examine the significance, if any, of this effect have been made by (1) isolating β-NGF from the complex by different methods and (2) seeking modifications of β-NGF which do not alter its specific NGF activity but do affect its ability to interact with the other two types of subunits.

With respect to the former, two new methods have been used. In one, 7S NGF was dissociated at alkaline pH and the β subunit isolated by ion exchange chromatography on QAE Sephadex at this pH. In contrast to the previous method utilizing ion exchange chromatography at acid pH, the β subunit does not bind to the column material and is eluted rapidly and quantitatively. In the second method, 7S NGF was subjected to isoelectric focusing in acrylamide gel and that part of the gel containing the separated β subunit cut out and eluted. In both instances, the physical and biological properties of these β-NGF preparations were identical to those of the original material.

Although limited exposure of β-NGF to high pH has no effect, prolonged exposure results in a significant increase in the minor component of lower isoelectric point, β_2, normally seen in β-NGF preparations. Continued exposure also produces a third species, β_3, of even lower isoelectric point. The proportion of the major β-NGF, β_1, decrease with time under these conditions. The interest in these new species lies in the fact that their ability to recombine with the α and γ subunits is much impaired. Under conditions, for example, where 100 % of β_1 recombines with appropriate proportions of α and γ subunits, only 20 % of β_2 will form the usual αβγ complex. Thus it is possible to use the β_2 and β_3 species to see if their specific biological activity is altered by the presence of α and γ subunits with which they effectively cannot interact. The result for β_2 showed that there was no difference in specific biological activity in the presence or absence of the two other subunits. The α and γ subunits do not, therefore, produce merely stabilization of β-NGF activity, rather is the specific interaction of the three subunits a factor in determining the level of the specific activity of β-NGF.

The Structure of β-NGF and a Comparison with 2.5 S NGF

β-NGF is the subunit which is isolated from the purified 7S NGF complex. The species now called 2.5 S NGF (2 and this volume) is isolated from a G-100 Sephadex filtrate of a streptomycin sulfate treated submaxillary gland extract (4) and, therefore, also originates from the 7S NGF complex. The only differences in the isolation of these two low molecular weight NGF species are the use of streptomycin sulfate and the stage at which the 7S NGF is dissociated and subjected to ion exchange chromatography at acid pH to isolate the smaller NGF protein. It is not surprising therefore that the two species are very similar. On the other hand, such differences that have been found are of very considerable interest.

There appears to be a difference in the amount of NGF activity which is recovered in the two forms. In a typical preparation, 7S NGF accounts for about 60 % of the activity measurable in the gland extract and β-NGF for about one fifth of this value (21). In contrast, some 60 % of the total NGF activity may be recovered as 2.5S NGF (4). The other difference is found at the structural level.The species, 2.5 S NGF, has two peptide chains, A, 118 residues long, and B, 110 residues long, whose amino acid sequences have now been determined (2, and this volume). Each contains three intra-chain disulfide bonds and the absence of sulfhydryl groups or of inter-chain disulfide bonds points to the strong non-covalent forces holding the two chains together. The sequence of the two chains is identical except at their N-termini where the B chain is missing the first eight residues of the A chain. The chains have C-terminal arginine residues and the comments already made about this property of β-NGF also hold for 2.5 S NGF. The sequence data obtained on β-NGF thus far show that it contains the same soluble tryptic peptides as 2.5S NGF which account, although their order is not completely established, for residues 51 through to the C terminus of either A or B chain. Only one major N-terminal sequence has been found for residues 1 through 34 and this corresponds to the sequence of the A chain. Since the amino acid composition of β-NGF is close to that of 2.5 S NGF (22) and a number of thermolytic peptides have been found which in composition cover most of the gap between residue 35 and 50 of the A chain, it may be tentatively concluded that β-NGF contains two identical A chains. The isoelectric focusing analysis of the reduced and carboxymethylated NGF proteins support this conclusion since β-NGF shows only one major component while 2.5 S NGF shows two in roughly equal amounts, one of which corresponds to the β-NGF component. Thus the NGF protein isolated from purified 7S NGF, i.e. the β subunit, has the composition AA and that isolated from 7S NGF in admixture with other gland proteins, i.e. 2.5 S NGF, the composition AB. Whether the loss of the octapeptide from the second A chain of 2.5 S NGF, which must occur during the dialysis of the G-100 filtrate to acid pH prior to chromatography, is an artefact or not is not known. If it represents the formation of the physiologically active peptide NH_2 ser.ser.thr.his.pro.val.phe.his COOH by specific enzymatic cleavage of the A chain, then it would be a significant pointer as to why such large amounts of the NGF protein are found in the adult mouse. Answers to this question may be obtained by examining the potential pharmacologic activity of the peptide and also by isolating the enzyme responsible for its production and determining its specificity. It will also be of interest to determine if the difference in structure between β and 2.5 S NGF accounts for the apparent differences in biological activity yields or if it has any effect on the interaction of these proteins with the α and γ subunits.

ACKNOWLEDGEMENTS

J.R. Perez-Polo : post-doctoral fellow, National Multiple Sclerosis Society. W.W.W. de Jong : postdoctoral fellow, UNESCO International Brain Research Organization. Present Address : University of Nijmegen, Nijmegen, The Netherlands.

Research described in this paper was supported by research grants from the National Institutes of Health (NS 04270) and the National Science Foundation. We are greatly indebted to Drs. Ruth Angeletti and Ralph Bradshaw for discussing their own work with us prior to publication.

REFERENCES

1 ANGELETTI, P.U., GANDINI-ATTARDI, D., TOSCHI, G., SALVI, M.I. and LEVI-MONTALCINI, R., Biochim. Biophys. Acta, 95 (1965) 111.
2 ANGELETTI, R.A.H. and BRADSHAW, R.A., Proc. Natl. Acad. Sci.US, 68 (1971) 2417.
3 ATTARDI; D.A.; SCHLESINGER, M.J. and SCHLESINGER, S., Science, 156 (1967) 1253.
4 BOCCHINI, V. and ANGELETTI, P.U., Proc. Natl. Acad. Sci. US, 64 (1969) 787.
5 GREENE, L.A., SHOOTER, E.M. and VARON, S., Proc. Natl. Acad. Sci. US, 60 (1968) 1383.
6 GREENE, L.A., SHOOTER, E.M. and VARON, S., Biochemistry, 8 (1969) 3735.
7 GREENE, L.A., TOMITA, J.T. and VARON, S., Exptl. Cell Res., 64 (1971) 387.
8 GROSSMAN, A., LELE, K.P., SHELDON, I., SCHENKEIN, I. and LEVY,M., Exptl. Cell Res., 54 (1969) 260.
9 HOFFMAN, H., in Biochemistry of Brain and Behavior (Ed. BOWMAN, R.E. and DATTA, S.P.) Plenum Press, New York, 1970.
10 LEVI-MONTALCINI, R. and ANGELETTI, P.U. in Growth of the Nervous System (Ed. WOLSTENHOLME, G.E.W. and O'CONNOR, M.) Churchill, London, 1968.
11 LEVI-MONTALCINI, R., MEYER, H. and HAMBURGER, V., Cancer Res., 14 (1954) 49.
12 LIUZZI, A., POCCHIARI, F. and ANGELETTI, P.U., Brain Res., 7 (1968) 452.
13 PARTLOW, L.M. and LARRABEE, M.G., J. Neurochem., 18 (1971) 2101.
14 PEREZ-POLO, J.R., BAMBURG, J.R., DE JONG, W.W.W., STRAUS, D., BAKER, M. and SHOOTER, E.M., in Nerve Growth Factor and its Antiserum (Ed. ZAIMIS, E., VERNON, C.H. and EDWARDS, D.C.)

Athlone Press, London, 1972.
15 ROISEN, F.J., MURPHY, R.A., PICHICHERO, M.E. and BRADEN, W.G., Science, 175 (1972) 73.
16 SMITH, A.P., VARON, S. and SHOOTER, E.M., Biochemistry, 7 (1968) 3259.
17 TAYLOR, J.M., COHEN, S. and MITCHELL, W.M.,Proc. Natl. Acad. Sci. US, 67 (1970) 164.
18 THOENEN, H., ANGELETTI, P.U., LEVI-MONTALCINI, R. and KETTLER,R., Proc. Natl. Acad. Sci. US, 68 (1971) 1598.
19 TRESKA-CIESIELSKI, J., GOMBOS, G. and MORGAN, I.G., C.R.Acad. Sci.Paris, 273 (1971) 1041.
20 VARON, S., NOMURA, J. and SHOOTER, E.M., Biochemistry, 6 (1967) 2202.
21 VARON, S., NOMURA, J. and SHOOTER, E.M., Biochemistry, 7 (1968) 1296.
22 VARON, S. and SHOOTER, E.M. in Biochemistry of Brain and Behavior (Ed. BOWMAN, R.E. and DATTA, S.P.) Plenum Press, New York, 1970.

STRUCTURAL STUDIES OF 2.5 S MOUSE SUBMAXILLARY GLAND NERVE GROWTH FACTOR

Ruth HOGUE ANGELETTI†, William A. FRAZIER and Ralph A. BRADSHAW
Department of Biology† and Department of Biochemistry, Washington University, St. Louis, Missouri 63110 (USA)

During the more than twenty years since its discovery (11), the nerve growth factor (NGF) has been indeed demonstrated to be an essential protagonist in the growth and development of the sympathetic nervous system (12). The NGF-stimulated growth response has been amply described from both metabolic and ultrastructural viewpoints.

In more recent years, several laboratories have focussed their attention on the relation of these biological properties to the chemical structure of the NGF molecules. In particular, studies have centered on NGF isolated from the male mouse submaxillary gland because of its relative abundance and ease of purification (13). However, NGF from this source has been isolated in two different forms, which, while basically similar, also show apparent conflicting properties (16, 18). On the one hand, NGF can be isolated as a small basic protein with a sedimentation value of 2.5 S (5), while on the other, the nerve growth promoting activity has been isolated as a larger complex with a sedimentation value of 7S. In this latter case, dissociation results in the formation of three subunit types, designated α,β and γ (15). Although the function of the α and γ subunits in the physiological role of this factor remains to be rigorously established, it is clear that only the β subunit possesses NGF activity in the dissociated state. In addition, it has now been established that a high degree of similarity exists between the β subunit of 7 S NGF and 2.5 S NGF (SHOOTER, personal communication). Consequently, we have undertaken a detailed examination of the structure of 2.5 S NGF, the fundamental unit possessing NGF activity, as the first step in establishing the structure-function relationships of this factor.

Subunit Structure

Earlier observations of Zanini *et al.* (18) indicated that 2.5 S NGF, as prepared by the method of Bocchini and Angeletti (5), underwent dissociation to yield molecules that possessed only half the apparent molecular weight of the native molecule (28,000). However, these 14,000 molecular weight species were indistinguishable from native NGF with regard to both biological and immunochemical properties. Consequently, experiments to determine the number and kinds of polypeptide chains comprising 2.5 S NGF were carried out. Three experimental lines of evidence were obtained (3) and are summarized in Table I. In the first, sedimentation equilibrium analyses

TABLE I

The Subunit Structure of Mouse 2.5 S NGF[a]

Molecular weight by sedimentation equilibrium	Native NGF	28,973 ± 976
	Native NGF (6M guanidine HCl)	16,228 ± 585
	S-Carboxymethyl NGF (6M guanidine HCl)	14,530 ± 1,432
Amino terminal analyses[b]	Residues/29,000 MW:	SER : 2.35 MET[c]: 0.42
Tryptic peptide mapping	Residues lysine + arginine/29,000 MW:	34
	Residues lysine + arginine/14,500 MW:	16
	Tryptic peptides found :	15

[a]Data taken from Angeletti *et al.* (3).
[b]Determined by the method of Stark and Smyth (14).
[c]Sum of methionine and methionine sulfoxide. Values for methionine sulfoxide were uncorrected.

performed in the analytical ultracentrifuge, indicated that whereas native NGF possessed a molecular weight of 29,000, in good agreement with previous determinations (5), treatment of native NGF,with 6 M guanidine-HCl resulted in an apparent dissociation, yielding subunits approximately half as large. Reduction and carboxymethylation of disulfide bridges (4), did not appreciably lower the molecular weight further. Second, quantitative amino terminal analysis, by the method of Stark and Smyth (14), indicated two residues of serine and somewhat lesser amounts of methionine, after hydrolysis of the corresponding hydantoins. However, the significance of these results was diminished by the large correction factor which must be applied when serine is present as an amino terminal residue. Third, peptide maps of the soluble tryptic peptides of NGF that had been reduced and carboxymethylated with ^{14}C-iodoacetate indicated only 14 major tryptic peptides or about half the number expected from the lysine and arginine content of the 29,000 molecular weight protein. Additional confirmation of this data was obtained by the observation that only 5 residues of radioactively labelled half-cystine were recovered in these fractions. A sixth unique residue was isolated from the tryptic insoluble material after an additional thermolytic digestion. This accounting may be contrasted with the value of 12 residues of half-cystine calculated for native NGF.

The sum of these data are consistent only with a subunit model for NGF composed of two identical or very similar polypeptide chains that are associated by non-covalent forces. It is noteworthy that similar conclusions have been reported by Greene et al. (8) for the β-subunit of 7 S NGF.

Amino Acid Sequence

The complete amino acid sequence of 2.5 S NGF was determined from the structural analyses of peptides derived by digestion with trypsin, chymotrypsin, thermolysin and pepsin. The resulting mixtures of peptides from these digests were fractionated on columns of Dowex 50 X 8 and Dowex 1 X 2, and SE-Sephadex utilizing pyridine acetate gradients previously described (7). The purified peptides were structured by a combination of Edman procedures (6, 9, 17). The sum of these data allowed the elucidation of a single, continuous sequence of amino acids as shown in Fig. 1 (2). The tentative structure possesses amino terminal serine, in agreement with the amino end group analyses (vide supra) and carboxyl terminal arginine. Confirmation of this assignment was obtained from carboxypeptidase B digestion of S-carboxymethyl NGF. Further digestion by carboxypeptidase A released the amino acids expected in concordance with the proposed sequence.

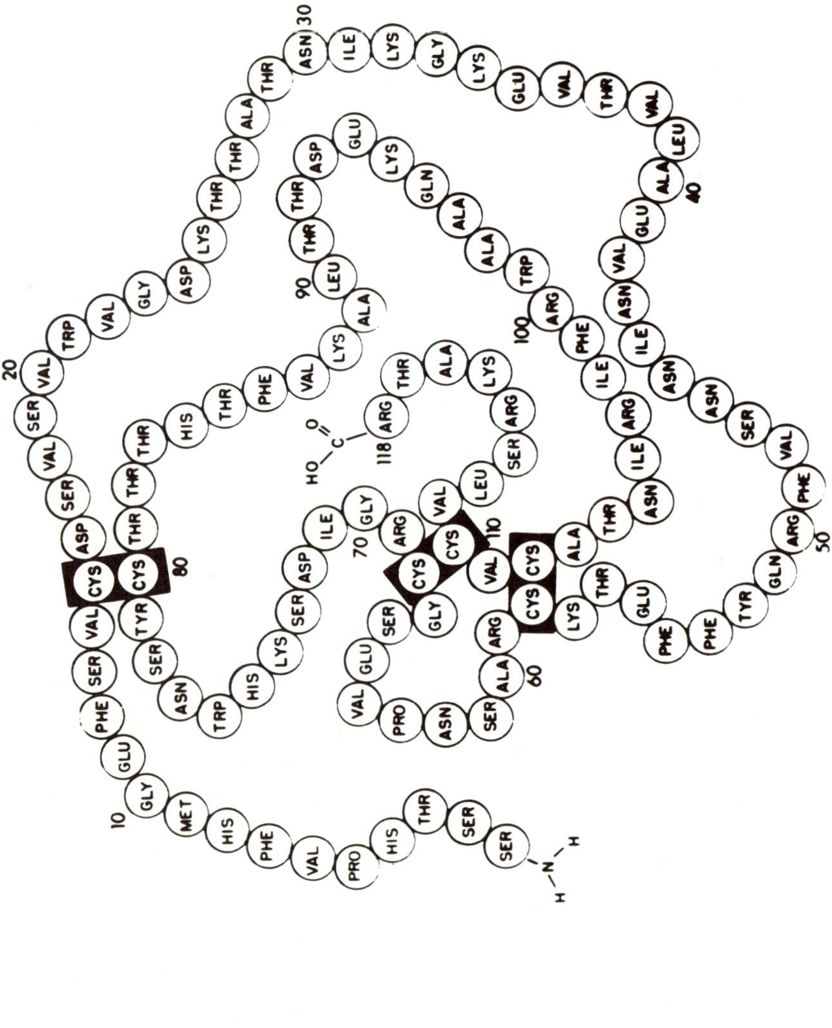

Fig. 1. Schematic representation of the amino acid sequence of the primary subunit of 2.5 S nerve growth factor from mouse submaxillary glands.

In addition to the alignment of the overlapping peptides used to generate the primary structure, it was necessary to determine the pairing of the half-cystinyl residues in order to complete the covalent structure of NGF. Accordingly, peptic digestion of native NGF was performed. The resulting peptides were separated on an SE-Sephadex column and all three disulfide bridges were found to be contained in one large peptide. This result is not surprising in view of the fact that two of the half-cystinyl residues, which were present in the single tryptic peptide are separated by a single valine residue (Fig. 1), producing a linear structure that would not be readily susceptible to attack by pepsin. Thus, further digestion with thermolysin, which readily cleaves at valine residues, was performed. After ion-exchange chromatography of this second hydrolysate, three distinct disulfide peptides were separated. These peptides were oxidized with performic acid, and the resultant half-peptides, listed in Table II, were purified by column chromatography or high voltage paper electrophoresis. The composition of each of the peptide pairs shown is unique, and allow the unambiguous assignment of the disulfide pairs in the NGF-molecule as being I-IV, II-V, and III-VI.

TABLE II

The Amino Acid Composition of Oxidized Disulfide Peptides Isolated from a Peptic/Thermolysin Digest of Native Nerve Growth Factor

Peptide	Composition	Residue No.
Pp XXIII-XXIV-Th 9-P1-H4	SER-VAL-CYS-ASP-SER	15 (CYS I)
Pp XXIII-XXIV-Th 9-P1-H5	TYR-CYS-THR	80 (CYS IV)
Pp XXIII-XXIV-Th 14-P2	VAL-CYS	58 (CYS II)
Pp XXIII-XXIV-Th 14-P5	SER-GLY-CYS-ARG-GLY	108 (CYS V)
Pp XXIII-XXIV-Th 16-17-P2	ALA-CYS	68 (CYS III)
Pp XXIII-XXIV-Th 16-17-P7	THR-LYS-CYS-ARG	110 (CYS VI)

Molecular Properties of 2.5 S NGF

The complete amino acid sequence of 2.5 S mouse NGF (Fig.1) is entirely consistent with the existing structural information for this protein (3-5). This polypeptide chain is composed of 118 amino acids with a molecular weight of 13,259 (2). Two such chains, when combined non-convalently in the dimeric structure of the native factor (3), have a molecular weight of 26,518. In this regard, it is noteworthy that information obtained from the amino acid sequenator, carried out in collaboration with Dr. Mark Hermodsen of the University of Washington, Seattle (10), indicates that NGF can be isolated in a form in which one of the two polypeptide chains comprising the native dimer is shorter, from the amino terminal end, by eight residues. The importance of these observations to the physiological role of NGF is presently under investigation.

In addition, several other features of the structure are distinctive. The amino terminal portion of the molecule is considerably less basic than the carboxyl terminal region. In fact, only three lysines and one arginine are found among the first 50 amino acids. This distribution is reflected in the fact that the tryptic insoluble core from S-carboxymethyl NGF, which is composed of two peptides covering residues 1-25 and 35-50, account for over 80 percent of these 50 residues. Conversely, the carboxyl-terminal portion of the molecule is more basic in character and, as judged by the alignment of the disulfide bonds, exists in a more rigidly confined conformation. The most striking feature revealed by the disulfide structure, when viewed within the context of the complete sequence,is the presence of a 14 residue loop that is formed by closing the disulfide bonds between half-cystines II-V and III-VI. The disulfide bonds apparently impart a particularly rigid and resistant nature to the NGF molecule, as indicated by its striking resistance to enzymatic, chemical and heat denaturation.

The electrostatic properties of the NGF are also explained by the assignment of the amide side chains. Seven of the eleven aspartate and two of the eight glutamate residues are present in the amide form. Thus, only ten of the nineteen potential acidic residues are present in the free acid form. This distribution of charge is in excellent agreement with the empirically determined isolectric point of 9.3 determined by isoelectric focussing.

The distribution of several other amino acid residues is also of interest. It may be readily noted that cyanogen bromide cleavage would not have been a productive approach for sequence analysis, for the single methionine residue is very close to the N-terminus, and is, in fact, the point at which the longer NGF chain is cleaved to yield the 8-residue shorter chain. Indeed, in this small region of the molecule are located one of the 2 prolines and two of the 4 histidine residues.

The completion of the primary structure of the NGF provides an opportunity to investigate those residues critical to its physiological action. Toward this end, studies have begun by carboxymethylating NGF at pH 6.8 with iodoacetic acid with the view of probing the relative importance of the individual histidine residues of this molecule. These studies have revealed the fact that there is not an obvious relationship between histidine residues and physiological activity. In fact, although these observations await confirmation with peptide isolation studies, it appears that 2 of the 4 histidine residues of NGF, which are located at positions 4, 8, 76 and 84, are freely available to modification without loss of biological properties. In a like manner, it has already been reported (1) that the three tryptophan residues of this protein located at positions 21, 77, and 99 can be subjected to oxidation via N-bromosuccinimide, resulting in derivatives that have partially lost activity. However, these modifications parallel losses in the tertiary structure of the protein, and suggest that the observed loss of biological activity is the result of general denaturation rather than the loss of residues associated with functional sites of the NGF. At present, it may be stated that the residues located in the primary structure of this protein responsible for the nerve-growth-promoting activity have not been defined.

Finally, it should be noted that NGF is only the second protein related to the nervous system for which complete sequence data is now available. Ultimately, it may be hoped that when additional structural features of other specific neural proteins are known, it will be possible to completely describe in molecular terms not only the relation of NGF to its role in stimulating differentiation in sympathetic neurons but other critical neural functions as well.

ACKNOWLEDGEMENTS

This investigation has been supported by grants from The National Institutes of Health, AM-13362 and NS-03777.

One of us (R.A.B.) has a research career development award from the National Institutes of Health, AM-23968.

The present address of Mrs. Angeletti is : Laboratorio di Biologia Cellulare, CNR, via G. Romagnosi 18/A, Roma, Italia.

REFERENCES

1 ANGELETTI, R.H., Biochim. Biophys. Acta, 214 (1970) 478

2 ANGELETTI, R.H. and BRADSHAW, R.A., Proc. Natl. Acad. Sci.US, in press.

3 ANGELETTI, R.H., BRADSHAW, R.A. and WADE, R.D., Biochemistry, 10 (1971) 463.

4 BOCCHINI, V., Europ. J. Biochem., 15 (1970) 127.

5 BOCCHINI, V. and ANGELETTI, P.U., Proc. Natl. Acad. Sci. US, 64 (1969) 787.

6 BRADSHAW, R.A., BABIN, D.R., NOMOTO, M., SRINIVASIN, N.G., ERICSSON, L.H., WALSH, K.A. and NEURATH, H., Biochemistry, 8 (1969) 3859.

7 BRADSHAW, R.A., ROBINSON, G.W., HASS, G.M. and HILL, R.L., J. Biol. Chem., 244 (1969) 1755.

8 GREENE, S.J., VARON, S., PILTCH, A. and SHOOTER, E.M.,Neurobiology, in press.

9 HARTLEY, B.S., Biochem. J., 119 (1970) 805.

10 HERMODSEN, M., ANGELETTI, R.H., FRAZIER, W.A. and BRADSHAW, R.A., manuscript in preparation.

11 LEVI-MONTALCINI, R., Ann. N.Y. Acad. Sci., 55 (1952) 330.

12 LEVI-MONTALCINI, R. and ANGELETTI, P.U., Physiol. Rev., 48 (1968) 534.

13 LEVI-MONTALCINI, R. and COHEN, S., Ann. N.Y. Acad. Sci., 85 (1960) 324.

14 STARK, G.R. and SMYTH, D.G., J. Biol. Chem., 238 (1963) 214.

15 VARON, S., NOMURA, J. and SHOOTER, E.M., Biochemistry, 6 (1967) 2202.

16 VARON, S., NOMURA, J. and SHOOTER, E.M., Proc. Natl. Acad. Sci. US, 57 (1967) 1782.

17 WATERFIELD, M. and HABER, E., Biochemistry, 9 (1970) 832.

18 ZANINI, A., ANGELETTI, P. and LEVI-MONTALCINI, R., Proc.Natl. Acad. Sci. US, 61 (1968) 835.

SECTION 3

CEREBRAL GLYCOPROTEINS

BIOCHEMISTRY, FUNCTION, AND NEUROPATHOLOGY OF THE GLYCOPROTEINS IN BRAIN TISSUE

Eric G. BRUNNGRABER

Research Department, Illinois State Psychiatric
Institute - Chicago, Illinois 60612 (USA)

Two reviews that summarize our knowledge of neural glycoproteins have appeared (12, 14). In this review, we will summarize some of the more recent developments in this field of neurochemistry.

The isolation of glycoproteins from brain is difficult since only 20 to 25 percent of the glycoprotein material can be rendered soluble by drastic homogenization in aqueous solvents. Furthermore, the isolation of purified glycoproteins presents technical problems due to their inherent microheterogeneity and physical properties. The task of establishing a quantitative method for the isolation of all of the glycoproteins of brain tissue is indeed a formidable task. If one will accept the hypothesis that the carbohydrate chains of the glycoproteins play important roles in brain development and function (6, 12-14, 16, 38) or in neuropathology (15), development of experimental procedures to isolate the heteropolysaccharide chains of the glycoproteins by proteolytic digestion of the defatted brain tissue samples appears justified. Proteolytic digestion of the samples solubilizes nearly all of the protein-bound N-acetyl-neuraminic acid (NANA) and hexosamine content of the tissue. The glycopeptides can be recovered quantitatively since no irreversible losses can occur during the procedure. On the other hand, in contrast to working with pure glycoproteins as starting material, one is confronted with a formidable array of glycopeptides the complete separation of which may prove to be exceedingly difficult. Nevertheless, if it can be shown that one, or more, specific heteropolysaccharide chain is altered due to functional or pathological changes, it would be possible to limit one's efforts toward the isolation of the glycoprotein which contains the altered chain. Hence, the isolation of the heteropolysaccharide chains may lead to information regarding function and pathology of the brain glycoproteins sooner than the more tedious development of methods for the quantitative separation and characterization of brain glycoproteins.

Isolation of Glycopeptides from Brain Glycoproteins

The procedure for the isolation of the glycopeptides has been described in detail elsewhere (19-23, 25, 27, 35). The defatted tissue sample is subjected to the proteolytic action of papain.The insoluble papain-resistant material is retrieved by centrifugation. The supernatant is dialyzed under standardized conditions. The non-dialyzable glycopeptide material is treated with cetylpyridinium chloride to remove by precipitation the glycosaminoglycans and nucleic acids. Excess cetylpyridinium chloride is removed by extraction with amyl alcohol. The non-dialyzable glycopeptides are separated from impurities which absorb ultra-violet light by subjecting the preparation to gel filtration on Sephadex G-50, using water as eluant (27). The dialyzable glycopeptides are recovered from the dialyzate by concentration and gel filtration on Sephadex G-15 to remove most of the amino acids, peptides, salts, and other impurities (35).

TABLE I

NANA- and Hexosamine-Containing Constituents of Rat Whole Brain

Constituent	NANA		Hexosamine	
	μM/g	%	μM/g	%
Gangliosides	2.22	67.5	0.96	23
Non-dialyzable glycopeptides	0.64	19.4	1.43	34
Dialyzable glycopeptides	0.23	7.1	0.81	19
Papain-resistant fraction	0.08	2.4	0.32	8
Free NANA	0.12	3.6	-	-
Free Hexosamine	-	-	0.14	3
Glycosaminoglycans	-	-	0.55	13
Total :	3.29	-	4.21	-

The distribution of various NANA- and hexosamine-containing materials is shown in Table I. Hexosamine in the glycopeptides accounts for 53 percent of the total content of hexosamine in rat brain. Some NANA and hexosamine invariably remains associated with the papain-resistant fraction from rat brain. This fraction from bovine brain contains little NANA or hexosamine (23) and that from human brain contains no NANA (25).

Chloroform-methanol extracts prepared from rat brain do not contain glycoprotein material of the type known to be present in the defatted tissue residue (25). Approximately 95 percent of the NANA in the extracts can be accounted for as ganglioside NANA (97). Such extracts do not contain glucosamine, mannose, or fucose.These sugars are prominant constituents of the rat brain glycoproteins.

The Non-Dialyzable Glycopeptides

The ratio of aspartic acid/threonine did not change after incubation of the non-dialyzable glycopeptides for 17 hours at 36° in 0.1 N NaOH, indicating that the linkage of the heteropolysaccharide chains to the protein is predominantly the alkali-stable β-aspartylglycosaminylamine linkage. Similar results were reported by Arima et al. (2). However, the glycopeptide mixture does contain a minor fraction that is susceptible to cleavage by alkaline treatment. This was noted by subjecting the alkali-treated non-dialyzable glycopeptide preparation to gel filtration on standardized columns of Sephadex G-50. The molecular weight of most of the NANA and hexose associated with the non-dialyzable glycopeptides was not affected. However, approximately 11-13 % of the total NANA appeared in a polysaccharide of decreased molecular weight. Presumably,this material formed as a consequence of the cleavage of a galactosamine-threonine (serine) linkage. This material may account for the small amount of galactosamine in the non-dialyzable glycopeptide preparation. Approximately 90-95 percent of the hexosamine consists of glucosamine ; the remainder is galactosamine.

The non-dialyzable glycopeptides have been subjected to column electrophoresis (Fig.1). Upon re-electrophoresis, the glycopeptides of fractions I through VI migrate to the same position. Contents of test tubes containing material corresponding to fractions I through VI (Fig. 1, upper left) were combined and analyzed for sugar components (Table II). Each fraction was subjected to gel filtration on Sephadex G-50 (Fig. 2) in order to estimate its average molecular weight. Glycopeptides of known molecular weight (60,93) were used to standardize the Sephadex G-50 columns.

The molecular weight of the non-dialyzable glycopeptide fractions (Table II) decreases as the charge (electrophoretic mobility) decreases. The glycopeptide fractions appear to differ in molecular

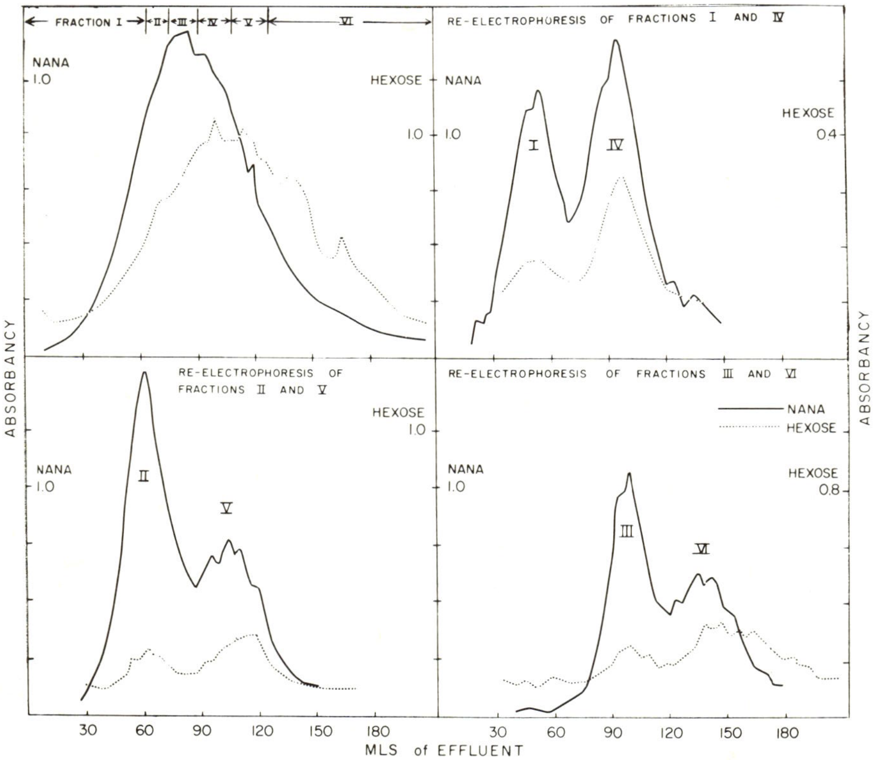

Fig. 1. Electropherogram obtained when rat brain non-dialyzable glycopeptides were subjected to column electrophoresis utilizing the LKB apparatus n° 3340. Electrophoresis was conducted at 3° in 0.1 M glycine-NaOH buffer, pH 10.3, at 200 V and 25-27 mA for 55-60 hours. The column was eluted with the same buffer and 3 ml fractions were collected. The fractions were analyzed for NANA and hexose. Contents of test tubes (graph at upper left) containing material corresponding to fractions I through VI were combined, dialyzed, concentrated, and analyzed for carbohydrate constituents and molecular weight (see Table II). The six fractions were also subjected to re-electrophoresis and the results are depicted in the figure.

TABLE II

Carbohydrate Composition and Average Molecular Weight of Non-Dialyzable Glycopeptides Obtained by Column Electrophoresis

Fraction	Molar ratios Mannose	Galactose	N-acetyl glucosamine	NANA	Fucose	Average MW	Average MW after mild hydrolysis	Loss in MW after mild acid hydrolysis
I	1	2.3	2.8	1.7	0.43	5200	3650	1550
II	1	2.3	3.4	2.1	0.43	4950	3500	1450
III	1	1.4	1.7	1.1	0.36	4850	3650	1200
IV	1	1.1	2.1	0.9	0.46	4800	3750	1050
V	1	1.0	1.8	0.6	0.50	4550	3450	1100
VI	1	0.8	1.5	0.3	0.52	4300	3650	650
Totals :	1	1.2	1.9	0.8	0.47			

Contents of test tubes corresponding to fractions I through VI (see Fig.1) were combined and analyzed for carbohydrate components. Each fraction was subjected to gel filtration on standardized columns of Sephadex G-50 in order to estimate its average molecular weight (see Fig.2). Each fraction was subjected to mild acid hydrolysis (0.1 N sulfuric acid, 80°, 1 hour) for removal of NANA, and the molecular weight of the NANA-free glycopeptides was estimated.

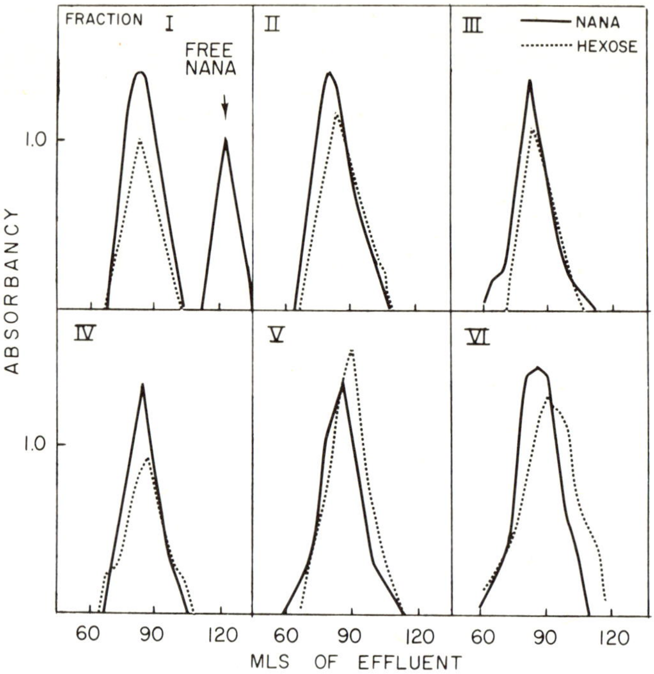

Fig.2. Chromatogram obtained when fractions I through VI (see Fig. 1) were subjected to gel filtration on standardized columns of Sephadex G-50 (1 x 90 cm), in 0.05 M Tris-chloride buffer, pH 7.5, containing 0.1 M NaCl. Columns were previously standardized with glycopeptides of known molecular weight (glycopeptide B from thyroglobulin and glycopeptide from transferrin). The position of free NANA is shown in figure at top, left hand corner.

weight largely because of differences in their content of NANA groups. If one assumes that fractions I and II contain 2 molecules of mannose, then each glycopeptide fraction would contain 4 molecules of NANA. The loss of these terminal NANA residues as a consequence of the mild acid hydrolysis would decrease the molecular weight of fractions I and II by 1240. If one assumes that the glycopeptides of fractions III through VI contain 3 molecules of mannose, then glycopeptides of fraction III, IV, V, and VI would respectively contain 3, 3, 2, and 1 molecules of NANA. The loss in molecular weight after mild acid hydrolysis would be 930, 930, 620, and 310 respectively for fractions III, IV, V, and VI. The decrease in molecular weight observed experimentally is somewhat greater than these values. This can be explained by the finding that the mild acid hydrolysis procedure known to cleave all of the NANA residues also cleaves approximately 8 percent of the hexose and 4 percent of the fucose residues. Approximately 5 percent of the hexosamine is also liberated (57).

It should be recognized that each of the six fractions consists of a mixture of glycopeptides which are very similar in molecular weight and electrophoretic mobility, and that the values for their sugar composition and molecular weight are average values for the mixture.

The non-dialyzable glycopeptides contain sulfate-ester groups (2, 25, 68). The preparation does not contain glycosaminoglycans since glucuronic acid is absent (2, 19, 20, 22) and their electrophoretic mobility exceeds that of the sulfated glycopeptides (25). Removal of NANA by mild acid hydrolysis, followed by subjection of the de-sialidized glycopeptides to column electrophoresis, resulted in the separation of sulfated and non-sulfated glycopeptides. Approximately 50 percent of the hexose and hexosamine of the de-sialidated non-dialyzable glycopeptide preparation appeared to be present in sulfated glycopeptides. Sulfated glycopeptides which contain large amounts of mannose, fucose, and NANA are clearly more closely related in composition to the heteropolysaccharide chains of glycoproteins than to the classical keratin sulfate, which is a linear polymer of glucosamine and galactose (56, 69). Consequently, the use of the term "keratosulfate" to describe these glycopeptides should be discontinued.

The non-dialyzable glycopeptide preparation was subjected to mild acid hydrolysis in 0.05 M HCl at 100° and the monosaccharides released at various time intervals were isolated by gel filtration on Sephadex G-15. All of the NANA is released in one hour. 75 percent of the fucose is liberated by 4 hours. An immediate release of galactose occurs so that approximately one-half is released as a monosaccharide by 6 hours, suggesting the galactose is located in an external position. Mannose is not liberated as a monosaccharide until 7 hours of hydrolysis ; it requires 8 to 12 hours before half of the mannose has appeared as a free sugar. Mannose, therefore, appears to be located in an internal position. Hexosamine release showed a slight lag period in the early phase of hydrolysis, with a maximal rate of release occurring between 6 and 8 hours of hydrolysis. This suggested that hexosamine appears in both external and internal positions in the heteropolysaccharide chains. The order of release was : NANA, fucose, galactose, glucosamine, mannose, glucosamine. The results obtained are similar to those found by Spiro (91) who followed the release of carbohydrates from the heteropolysaccharide chains of fetuin. The order of release of monosaccharides is therefore consistent with the hypothesis that the bulk of the glycopeptides consist of heteropolysaccharide chains that are quite similar to those isolated from other sources (see structures below). The presence of sulfate ester groups clearly distinguishes the brain non-dialyzable glycopeptides from glycopeptides derived from glycoproteins such as fetuin (91), orosomucoid(102) and α_2-macroglobulin (41, 94). Sulfate ester groups appear to be located in external positions (linked to galactose) as well as internal positions (linked to N-acetylglucosamine (68).

Dialyzable Glycopeptides

The dialyzable glycopeptides were subjected to column electrophoresis (Fig. 3). Contents of test tubes containing fractions I through VII were combined and subjected to gel filtration on standardized columns of Sephadex G-50. Unlike the chromatographic profiles obtained for the non-dialyzable glycopeptides (Fig.2), the various fractions exhibited considerable heterogeneity upon gel filtration (Fig. 4). Several fractions (IA, IIA, IIIA, IVA, and V) had properties that were identical to glycopeptide fractions which appear in the non-dialyzable glycopeptide preparation. For example, fractions IA, IIA, IIIA, IVA, and V (Fig. 4) have molecular weights and carbohydrate compositions that are similar to those of fractions I, II-III, III-IV, V, and VI recovered by column electrophoresis of the non-dialyzable glycopeptide preparation (Fig. 1, Table II). These do not contain glucose. Glucose appears to be a constituent sugar only of the dialysable glycopeptides (IB, IIB, IIC, IIIC, and possibly VI and VII). The presence of glucose-containing glyco-

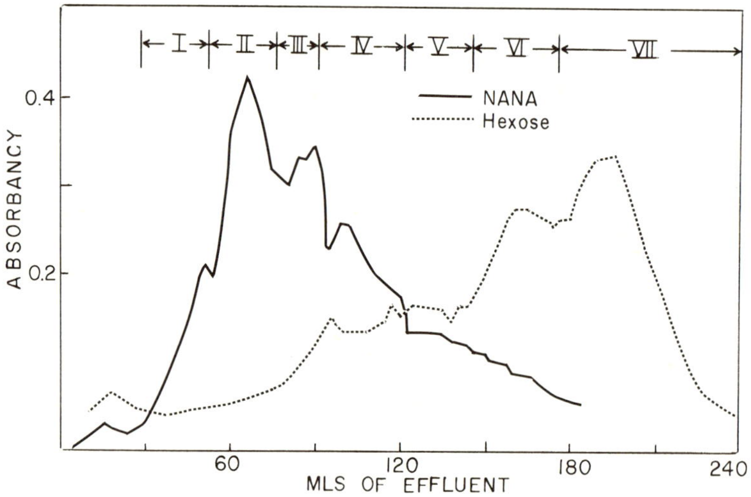

Fig. 3. Electrophoregram obtained when rat brain dialyzable glycopeptides were subjected to column electrophoresis. Conditions of electrophoresis were identical to those described in Fig. 1. Contents of test tubes containing material corresponding to fractions I through VII (see top of graph) were combined, concentrated, desalted by gel filtration, and subjected to gel filtration on standardized columns of Sephadex G-50 (see Fig. 4).

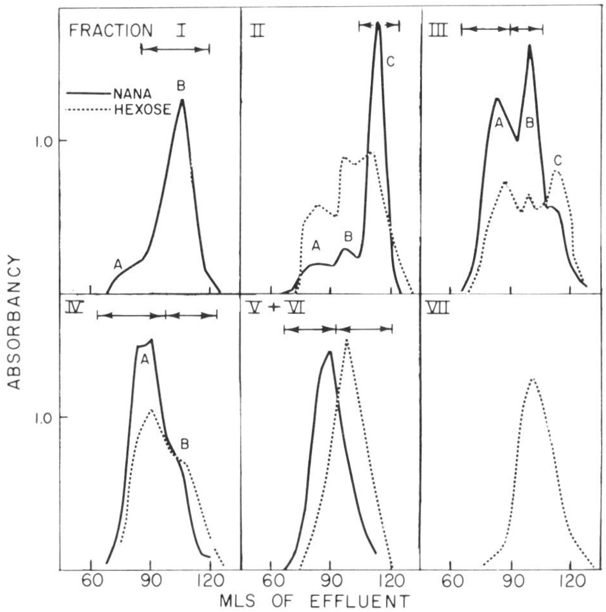

Fig. 4. Chromatogram obtained when fractions I through VII (see Fig.3) were subjected to gel filtration on standardized columns of Sephadex G-50. Conditions are the same as those described under Fig. 2.

peptides is of interest in view of the report of Riddell and Leonard (80),who provided evidence that glucose-containing glycopeptides derived from brain glycoproteins cause coma upon injection into experimental animals. The hexosamine content of fractions IB, IIC, IIIB, and IIIC is partly or predominantly galactosamine. Glucosamine is the sole hexosamine in fractions IIIA, IVA, IVB, V, VI and VII. Some of the dialyzable glycopeptides may contain sulfate-ester groups (IIB, IIIB, and IVB).

Dialyzable glycopeptides of fractions VI and VII differ markedly from other glycopeptides examined because of their high content of mannose, absence of NANA, and very low levels (if any) of fucose. If one assumes that the glycopeptides of fractions VI and VII are 70 percent in carbohydrate content, the heteropolysaccharide chain would have a molecular weight of 2129. This would accomodate 8 mannose and 4 N-acetylglucosamine residues. One fucose and one galactose residue may also be present. The behaviour of the hexosamine in these glycopeptides in certain colorimetric methods used to distinguish between glucosamine and galactosamine led to the suggestion that the hexosamine may be mannosamine (18). The use of the Gardell column in combination with spectrophotometric methods (18) have indicated that the hexosamine is exclusively glucosamine (25).

Subcellular Distribution of Glycoproteins

Published investigations on the subcellular distribution of glycoproteins (20, 26, 31, 32) have been extensively reviewed (14). The synaptosomal and microsomal preparations from brain account for most of the glycoprotein material of the tissue. However, it has been shown that the usual synaptosomal preparation is contaminated with structures that are morphologically similar to axonal and glial fragments (31, 32) and these may contain glycoproteins. On the other hand, Fisher et al. (46) have found that isolated axolemma from the retinal axons of the squid contain less NANA than that found in liver plasma membranes. The mitochondrial fractions contain very little glycoprotein material, part of which is probably associated with the lysosomes (63) which sediment in this fraction. Lysosomes contain several acid hydrolysases which are glycoproteins, although the lysosomal membrane does not appear to contain NANA (88). Purified brain mitochondrial preparations contain very little, if any, glycoproteins, as judged by their NANA content (Marks, personal communication). The crude nuclear preparation accounts for only 20 percent or less of the total brain glycoproteins, some of it undoubtedly associated with red blood cells and blood vessels which sediment in this fraction. The presence of glucosamine -containing material in purified nuclear membranes from rat liver has been reported (62). Castejon (29) reported the presence of NANA on the cell surfaces of neurons.

Anatomical Distribution of Glycoproteins

Brain tissue samples that consist largely of cell bodies contain a higher concentration of glycoproteins and gangliosides than areas that are enriched in fiber tracts (23, 81). The concentration of glycoproteins relative to gangliosides is somewhat more elevated in white matter than in gray matter. Since the concentration of glycoproteins relative to gangliosides is greater in the synaptosomal fraction than in axonal fragments, it was suggested (14, 23) that oligodendroglial cells in white matter contain glycoproteins and that fragments originating from these cells contaminate the usual synaptosomal fraction. This was in agreement with morphological evidence (31, 32). Arseni and Carp (3) have reported the synthesis of glycoproteins in astrocytomas and multiform glioblastomas.

Glycoproteins in Development

A large part of the non-dialyzable glycopeptide content, per gram rat brain, was synthesized prior to the acceleration of ganglioside deposition which occurs between 6 and 29 days of age (57). There is a slow and steady increase in glycopeptide deposition between 6 and 29 days of age. Quarles and Brady (76) studied total glycoprotein-NANA content and obtained similar results. Nearly adult levels are reached by 29 days of age. The composition of the non-dialyzable glycopeptides derived from glycoproteins from the 15-day old rat showed qualitative differences compared to those isolated from the adult brain (57). Di Benedetta et al. (37) found a continuous heterogenization of soluble glycoproteins in the neonatal rat during the first 20 days of life, as judged by electrophoretic experiments. Some glycoproteins soluble in solvents of low ionic strength increased up to 15 days of life and then quickly decreased. Two glycoprotein fractions that became extensively labelled after intracerebral injection of fucose were found in the 1 to 10 day old mouse. These were not prominent in the 15-day old animal (43). Roukema et al. (82) reported a marked increase in glycoprotein content between 4 and 18 days of age in the rat.

Soluble Glycoproteins

The non-dialyzable glycopeptides from the soluble glycoproteins contain less fucose than those obtained from the insoluble residue. A greater percentage of the total glycopeptide material derived from the soluble glycoproteins is non-dialyzable. A column electrophoretic separation (36) of the soluble glycoproteins of rat brain (Fig. 5) and analysis of 6 fractions (Table III) showed that glycoprotein-carbohydrate was present in all fractions. However, fractions I and II showed an enrichment in carbohydrate content, per mg protein, compared to the original extract. Fraction I was shown to contain 4 glycoproteins which migrate more rapidly than serum or brain albumin upon electrophoresis in polyacrylamide gel. Three of these glycoproteins were not found in liver extracts or in serum. One or more of the acidic proteins described by McEwen and Hyden (70) appear to be glycoproteins. All of the pre-albumin proteins, including the four glycoproteins and the S-100 protein, appear in fraction I (Fig. 5) and are separated from the remaining protein material of the soluble extract. An interesting anomaly concerning the soluble glycoproteins was the finding that when lyophilized brain extracts are taken up in Tris-chloride buffer and subjected to gel filtration on Sepharose 4B, all of the glycoprotein material is excluded from the gel matrix. A three-fold enrichment in glycoprotein-carbohydrate, relative to the original, is obtained. Apparently, the glycoproteins can, under certain conditions, form aggregates with an apparent molecular weight exceeding two million. The true molecular weight for most of the glycoprotein material is considerably less than this value, since they are capable of entering the pores of a 7 percent acrylamide gel during electrophoresis in borate buffer, pH 9.2.

Several workers have reported the isolation of soluble brain specific glycoprotein antigens (54, 61, 66, 77, 90, 98, 103, 104). Lysosomal hydrolytic enzymes (63) and brain alkaline phosphatase

TABLE III

Analysis of Fractions I through VI Obtained by Preparative Column Electrophoresis of Rat Brain Soluble Glycoproteins (Fig.5)

Fraction	mg carbohydrate/mg protein	µg NANA/ mg protein	Percentage of total protein recovered in fraction	Molar ratios of carbohydrate constituents			
				NANA	Fucose	Hexose	Hexosamine
I	0.053	8.7	7.4	1	0.35	3.9	3.9
II	0.020	3.4	9.2	1	0.31	2.6	2.9
III	0.010	2.0	23.2	1	0.30	3.0	2.9
IV	0.009	1.6	18.6	1	0.30	3.2	3.3
V	0.010	1.2	19.4	1	0.50	6.2	3.7
VI	0.010	0.6	22.1	1	0.65	25.5	6.2
Original extract :	0.011	2.3					

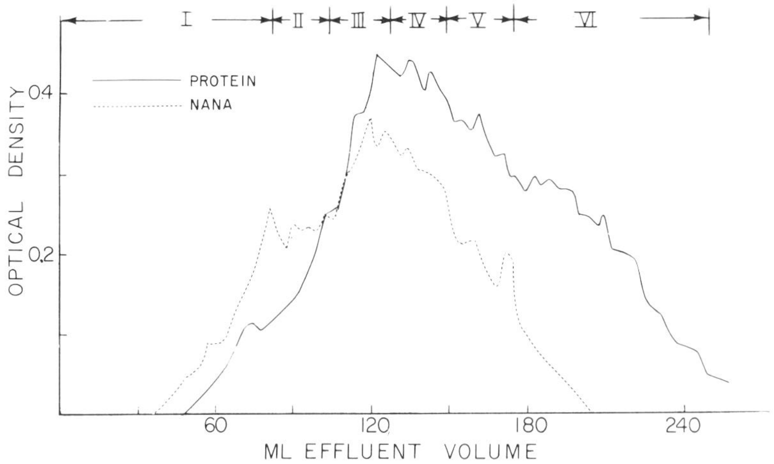

Fig. 5. Soluble proteins were extracted from tissue with 0.02 M Tris-HCl buffer, pH 8.0 by homogenization of whole rat brain tissue (1:5, w/v) with a Servall Omni-Mixer for 5 min at approximately 8,000 rev./min. The homogenate was centrifuged at 100,000 x g (average force) for 1 h. The residue was re-extracted with water (1:5, w/v) using the omnimixer and centrifuged as before. The combined supernatants were lyophilized. The dry material was resuspended in water and dialyzed against the buffer that was used for electrophoresis. Column electrophoresis was performed in 0.033 M sodium phosphate, pH 7.4. The anode is at the left. Contents of test tubes containing material corresponding to fractions I through VI were combined and analyzed for protein and carbohydrate constituents. Electrophoresis was conducted on the LKB n° 3340 column using formalated cellulose as support, at 200 V for 48 h.

(85) are known to be glycoproteins.

Wilson *et al.* (106) found that vinblastin precipitated a number of proteins in addition to neurotubular protein. These proteins are also precipitated by calcium ions. Dekirmenjian and Liu (33) have reported that glycoproteins of the soluble extracts are precipitated along with neurotubular proteins by vinblastin.

Insoluble Glycoproteins

The first attempt to dissolve and purify the insoluble glycoproteins from brain tissue was reported by Brunngraber *et al.* (17). The combined mitochondrial and microsomal preparation from rat brain was used as starting material. Only 10 percent of the particulate glycoprotein could be solubilized by extraction with aqueous buffer. Approximately 60 percent was solubilized by treatment with Triton

X-100. A 100 percent recovery of Triton-solubilized glycoproteins was obtained by fractionation on calcium hydroxylapatite columns. The amount of carbohydrate, per mg protein, of some fractions showed a three-fold enrichment relative to that of the original extract. Purification of glycoproteins can also be achieved by chromatography on DEAE-cellulose. In both procedures, all eluting solutions must contain the detergent, Triton X-100.

Procedures for solubilization of insoluble proteins by means of sodium dodecylsulfate, reduction by 2-mercaptoethanol, and subsequent electrophoretic separation in polyacrylamide gels in which the buffer contained the detergent has been described by Maizel and coworkers (67, 89). These procedures were used by Bosmann *et al.* (8) to separate glycoproteins from subcellular fractions from brain tissue. Unfortunately, glycolipids and lipids also stain positively with the periodic acid-Schiff staining procedure used in this work. Consequently, the conclusion drawn by these authors that 9 of their protein bands are glycoproteins may be premature. Dutton and Barondes (43) used similar procedures for the separation of glycoproteins that had been labelled *in vivo* with radioactive fucose. Label was found in all of the fractions eluted from the gel. Susz (96) varied the procedure by adding a prior delipidation step.More recently, Dr. Susz, working in our laboratory, was able to solubilize all of the glycoprotein material of the residue that is obtained after extraction of a crude mitochondrial fraction with chloroform-methanol (2:1 and 1:2, v/v). Solubilization was achieved by treatment of the delipidated material with EDTA, sodium dodecyl sulfate, and mercaptoethanol. An electrophoretic method recently described (50) utilizing porosity gradient polyacrylamide gels has been found successful in the separation of 7 or more glycoprotein bands from delipidated crude mitochondrial fractions.

Biosynthesis of Brain Glycoproteins

After intraperitoneal injection of D-glucosamine-1-^{14}C into the 15 day-old rat, maximal activity in the brain non-dialyzable glycopeptides was attained 4 hours after injection (57). The rate of decline in radioactivity after maximal incorporation suggested the presence of at least two components, the apparent half-lives of which were approximately 18 hours and 4 days. In contrast, maximal activity into the gangliosides was not attained until 24 hours after injection and the loss of label after maximal incorporation proceeded at a slower rate. In the adult rat, the incorporated activity reached nearly maximal values by 8 hours after injection. The decline in radioactivity from its peak value at 24 hours suggested the presence of two components which possessed half-lives of approximately 40 hours and 12.5 days respectively. Most of the radioactivity incorporated into the glycopeptides appeared in NANA and hexosamine ; very little label appears in the amino acids and hexose components of the glycopeptides.

It is well established that glycoprotein synthesis in the liver (86), HeLa-cell (9), and in a variety of cells (78) occurs in the smooth endoplasmic reticulum or Golgi apparatus. The protein is synthesized at the ribosomal level, and the protein molecule then passes through the channels of the Golgi apparatus. The internally located sugars, glucosamine and mannose, are added to the protein in the early portion of the passage, and galactose, NANA and fucose are added later. Droz (40) presented radioautographical evidence that this is the case in neurones ; no glycoprotein synthesis could be detected in the nerve-endings. Since there was a more rapid incorporation of labelled glucosamine into macromolecules of the synaptosome, as compared to incorporation of labelled amino acids, Barondes (4) suggested that glycoproteins are synthesized at this site. Den et al. (30) reported a high concentration of glycosyl transferases in synaptosome-enriched fractions. Forman et al. (47) found that labelled glucosamine that had been injected into the goldfish eye is incorporated into glycoproteins which are transported through the optic nerve to the contralateral optic tectum and that glycoproteins appear to be among the more rapidly migrating components. Elam et al. (44) noted that a sulfate-labelled material of electrophoretic mobility less than that of the glycosaminoglycans migrated along the optic nerve. Since the material was prepared by methods similar to our own, it appears that this material was identical to the non-dialyzable glycopeptides. These studies suggest that the early appearance of labelled glucosamine in the synaptosome may be due to synthesis in the cell body and rapid migration to the nerve endings. It should be recalled that the usual synaptosomal preparation contains axonal fragments (31, 32). Roux (84) has reported that synthesis of glycoproteins occurs in the supra-optic nucleus, and that these are subsequently transferred to the posterior pituitary.

Seijo and Rodriguez De Lores Arnaiz (87) used thiamine pyrophosphatase as a marker for the Golgi complex in order to determine its distribution in various subcellular fractions obtained by sucrose-density gradient centrifugation techniques. Most of the activity was found to sediment in the synaptosomal fraction as well as in synaptosomal membranes isolated from synaptosome-enriched fractions. The enzyme was also found in Golgi-like structures that accumulate at segments near the compression of crushed cat sciatic nerves (75).

Bosmann et al. (9) were able to show that the synthesis of the carbohydrate portion of the glycoproteins in the HeLa-cell occurs within the Golgi apparatus. Subcellular fractions enriched in plasma membrane fragments were inactive. They believed that the product of this synthesis is a complete glycoprotein molecule that is subsequently incorporated into plasma membranes by integration with lipids and other plasma membrane components.

At present, there is no convincing evidence that neuronal glycoproteins are synthesized at the site of the synaptosomal membrane. Experimental evidence for glycoprotein synthesis in the nerve-ending or its membranes requires proof that these fractions are free of axonal fragments (31, 32), glial fragments (32), and the Golgi apparatus (87).

Bosmann et al. (10, 11) have claimed that mitochondria can incorporate labelled hexose into glycoprotein molecules. The finding is difficult to reconcile with the evidence indicating that mitochondria do not contain glycoprotein material (see above). If mitochondria are indeed capable of synthesizing glycoproteins, it would appear that these glycoproteins are either synthesized for export, or alternatively, are deposited in amounts that are too low to detect by the usual chemical methods. In these series of experiments it was noted that asparaginase inhibited glycoprotein synthesis (7).

As far as is known, fucose in brain tissue is present only in the glycoproteins. Consequently, fucose is a good marker for glycoproteins in labelling experiments (5, 108). Some of the NANA that is injected intracerebrally becomes incorporated into rat brain glycoproteins (34).

Degradation of Glycoproteins

Little is known regarding the catabolism of glycoproteins. Glycoproteins can undoubtedly be broken down within the lysosomes which contain the required proteinases and acid hydrolases. Such catabolism would necessitate some mechanism whereby the glycoprotein on the cell surface is pulled back into the cell for catabolism. However, it is also possible that catabolism proceeds at the membrane surface, so that the heteropolysaccharide chains are cleaved from the protein and released into the intercellular spaces, the CSF, or the circulation. Goussault and Bourrillon (49) have reported that NANA-rich glycopeptides resembling those obtained from brain tissue appear in normal human urine.

Function of the Glycoproteins

It has been reported (6) that the amount and metabolism of glycoproteins in the pigeon brain are affected by training. These experiments have been criticized on biochemical (14) and behavioural grounds (48). We have attempted (58) to determine whether the metabolism and levels of the non-dialyzable glycopeptides from rat brain are affected by stress and during the process of memory consolidation. Rats were trained for 3-5 min on a simple one trial passive avoidance of foot shock task (83). A three-minute retention test was run 24 hours after original learning. Glucosamine-1-^{14}C was injected intraperitoneally 15 min before original learning (experiments 1 and 2), before retention (experiment 3), or before 10 min of inescapable foot shock (experiment 4). Control animals were handled in the same fashion as experimental subjects except that the former never received footshock. Sacrifice occurred one hour (experiments 1, 3, and 4) or 24 hours (experiment 2) after injection. There were no significant differences between controls and experimental animals in the incorporation of isotope into the heteropolysaccharide chains isolated from brain glycoproteins. The amounts of NANA, hexose, hexosamine, and fucose in the heteropolysaccharide chains were also unaffected.

Irwin and Sampson (59) reported changes in the content of di- and trisialo-gangliosides as well as changes in ganglioside turnover in rat brain following behavioural stimulation. A possible role for glycoproteins was suggested by a lower specific activity of the residue obtained after extraction of the gangliosides 9 and 24 hours after injection of labelled glucosamine in rats that had been taught a swim-escape task.

It has been proposed (12-14, 16) that alterations in the structure of the heteropolysaccharide chains as a result of stimulation may influence the subsequent performance of the cell. It is pertinent that Whetsell and Bunge (105) reported that exposure of the sensory ganglion to ouabain caused the Golgi apparatus to swell. The effect was reversible. If such changes occur upon stimulation, it is conceivable that structural alterations within the Golgi apparatus would result in alterations in the type of heteropolysaccharide chains being synthesized at the time of stimulation.

Langley (65) postulated the presence of carboxylated polyanions at the site of the Ranvier node which might act as macromolecular-polycation (Na, Ca) complexes in events associated with the neural impulse. The role of intercellular mucosubstances in neural information processing has been referred to in a review by Adey (1).

It has been proposed that (12-14, 16) glycoproteins on the cell surface have a directing influence on the formation of nerve connections in the developing nervous system. Richmond et al. (79) have reported that the incorporation of labelled glucosamine into glycoproteins was related to the degree of aggregation of dissociated chick embryo neural retinal cells. Free glucosamine and puromycin inhibited incorporation as well as aggregation.

The effect of blocking cholinergic and adrenergic transmission on glycoproteins in receptor cells was studied in a number of investigations. A decrease of glucosaminidase in hypothalamic nuclei as a consequence of inhibition of transmission was noted by Pasqualino and Tessitore (74). Dische _et al._ (39) reported that the composition of the heteropolysaccharide chains of glycoproteins produced by the submaxillary gland was dependent on whether the gland was stimulated by sympathetic or parasympathetic innervation. Denervation of rat diaphragm muscle caused an increase in the incorporation of glucosamine into glycoproteins of the muscle tissue (45). The possible role of glycoproteins in the serotonin receptor has received additional study (99,100).

The presence of NANA in Reissner's fiber has been demonstrated (95) and it was suggested that this fiber may balance the concentration of biogenic amines in the CSF by a cation exchange process. Polysaccharides appear to undergo histochemical changes during excitation and inhibition of neurons (64). Heller (55) has reported on the possible mechanism of function of a glycoprotein, a visual pigment, which is part of the membrane of the retinal rod outer segments.

In another connection, Yasui _et al._ (107) have reported on the isolation of an active fragment from bovine brain which included the active site of the cell receptor for the Japanese encephalitis virus. The activity was destroyed by α-mannosidase.

An interesting approach to a study of function was reported by Mihailovic _et al._ (71). They prepared antigen from brain areas and formed the anti-caudate nucleus and anti-hippocampal antibodies. Injection of these antibodies caused changes in behavioural responses. Such work could be extended to a study of purified glycoprotein antigens.

Neuropathology

Neuropathological samples studied include cases of subacute sclerosing leukoencephalitis (SSLE) (24), neurolipidosis with failure of myelination (51), generalized lymphosarcoma (neurologically normal), galactosemia (53), late infantile amaurotic idiocy (Bielschowsky type) (52), and Tay-Sachs disease (28) (Table IV). The major defect in SSLE was an elevation of a dialyzable glycopeptide containing largely mannose and glucosamine, but which did not contain NANA. In late infantile amaurotic idiocy and galactosemia, the major defect was a marked decrease in those dialyzable glycopeptides which do not contain NANA. In the case of neurolipidosis with failure of myelination, all of the glycopeptide fractions showed marked alterations.

The Tay-Sachs case is perhaps of greatest interest since it was established that the tissues of this patient lacked hexosaminidase A. Although the gangliosidic NANA increased four-fold due

TABLE IV
Carbohydrate Composition of Non-Dialyzable and Dialyzable Glycopeptides from Normal and Pathological Human Brain Tissue
(All values are expressed in terms of μmoles/g fresh tissue)

	NANA	Fucose	Hexosamine	Hexose
NON-DIALYZABLE GLYCOPEPTIDES				
Gray Matter				
Normal 8 year-old	0.72	0.55	1.94	2.17
SSLE, 8 year-old	0.63	0.40	1.59	1.79
Neurolipidosis,with failure of myelination,4 year-old	0.31	-	1.33	-
Generalized lymphosarcoma, 3 year-old	0.70	0.47	2.32	1.77
Galactosemia, 25 year-old	0.60	0.29	1.20	1.85
Late infantile amaurotic idiocy (Bielschowsky), 11 year-old	0.44	-	0.83	1.44
Tay-Sachs disease, 20 month-old	0.77	0.49	1.81	2.18
White Matter				
Normal	0.46	0.24	1.10	1.30
SSLE	0.50	0.26	1.17	1.38
Neurolipidosis, with failure of myelination	0.32	-	1.35	-
Galactosemia	0.59	0.21	1.20	1.67
Late infantile amaurotic idiocy	0.46	-	1.14	1.17
Tay-Sachs disease	0.65	0.38	2.36	1.79
DIALYZABLE GLYCOPEPTIDES				
Gray Matter				
Normal	0.21	0.22	1.03	1.05
SSLE	0.24	0.25	1.49	1.48
Neurolipidosis,with failure of myelination	0.57	-	0.91	-
Generalized lymphosarcoma	0.24	0.27	1.65	1.33
Galactosemia	0.27	-	0.40	-
Late infantile amaurotic idiocy	0.25	-	0.28	-
Tay-Sachs disease	0.24	0.37	1.93	3.57
White Matter				
Normal	0.18	0.17	1.08	0.81
SSLE	0.13	0.20	1.17	1.02
Neurolipidosis with failure of myelination	0.37	-	0.75	-
Galactosemia	0.27	-	0.35	-
Late infantile amaurotic idiocy	0.14	-	0.25	-
Tay-Sachs disease	0.18	0.34	1.72	1.51

to the accumulation of the neuraminidase-resistant ganglioside GM_2, the concentration of NANA in the non-dialyzable and dialyzable glycopeptides remained unchanged. It is known that Tay-Sachs ganglioside becomes elevated in this disease due to the terminal location of N-acetylgalactosamine which cannot be cleaved due to the deficiency of hexosaminidase A. Gangliosidic NANA becomes elevated since the cleavage of NANA is dependent upon the prior removal of the terminal N-acetylgalactosamine. This structural feature appears to be absent in the glycopeptides. This agrees with the finding that all of the NANA in the glycopeptides is removed by neuraminidase, but 40 % of the NANA in the gangliosides is resistant to the action of this enzyme.

Van Hoof and Hers (101)and Durand et al. (42) described cases in which the tissues lacked the enzyme α-fucosidase. This deficiency was accompanied by an increase in the fucose content of the "mucopolysaccharide" fraction (101). Glycoproteins are the only brain constituent that are currently known to contain fucose. Consequently, the absence of α-fucosidase would affect only this constituent ; the catabolism of the glycosaminoglycans should remain unaffected. As far as is known,fucose residues are present solely in terminal positions of the heteropolysaccharide chains. Hence, in fucosidosis, the block in biological degradation occurs at an initial stage of the catabolic sequence. Chains which contain fucose in the terminal position would not be amenable to subsequent attack by β-galactosidase, β-hexosaminidase, and α-mannosidase. These enzymes are believed to act in sequence by attacking the non-reducing ends of the chains. In this disease, one would expect an increased level of polysaccharide chains which contain fucose in a terminal position. The fucose-galactose-N-acetylglucosamine-portion of the heteropolysaccharide molecule will accumulate ; the NANA-galactose N-acetylglucosamine-chains are degraded (15).

Structures for the brain non-dialyzable glycopeptides are probably similar to those reported for orosomucoid glycopeptides (102) or fetuin (92) (Fig.6).

Ockerman (72) reported a case of mannosidosis in which an oligosaccharide consisting of 5 residues of mannose and 1 residue of glucosamine accumulated. In cases of mannosidosis, catabolism of the heteropolysaccharide chains of the glycoproteins would proceed so that (42) neuraminidiase, α-fucosidase, β-galactosidase, and β-hexosaminidase would remove the externally located sugars, leaving the mannose-rich, glucosamine-containing internal portion of the glycoprotein intact. Degradation then ceases since the enzyme, α-mannosidase, is absent. The water-soluble mannose-rich fragment obtained by Ockerman is derived by the action of an enzyme which splits the bond between N-acetylglucosamine and asparagine. The oligosaccharide then accumulates since it cannot be further degraded (15). Ockerman's oligosaccharide may be derived from the

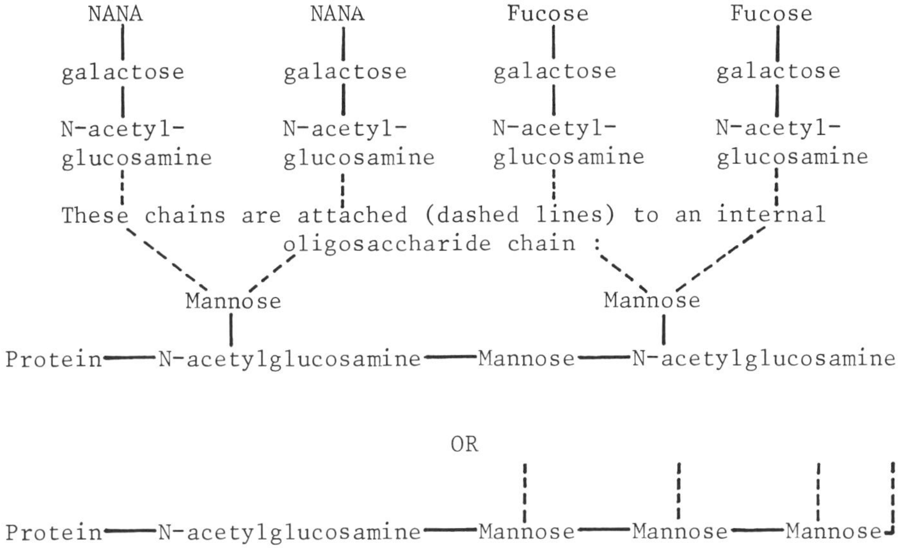

Fig.6. Possible structure for brain non-dialyzable glycopeptides.

catabolism of the non-dialyzable glycopeptides.However, the non-dialyzable glycopeptides from rat brain contain only two or three mannose residues per molecule of glycopeptide. The analytical values obtained for human brain suggests that this is probably the case in this tissue also. Consequently, it appears more likely that Ockerman's oligosaccharide is derived from the mannose-rich glycopeptide of the dialyzable fraction (fractions VI and VII, Fig.4). Oligosaccharide chains containing only two or three mannose residues are probably present in the cases of mannosidosis (73) and these are derived from the heteropolysaccharide chains present in the non-dialyzable glycopeptide fraction.

In the case of Tay-Sachs disease, the enzymatic cleavage proceeds so that the NANA,fucose, and galactose groups are cleaved.The N-acetylglucosamine which is attached to the mannose at a branch point may not be cleaved due to the deficiency of hexosaminidase A. A three-fold elevation in the concentration of hexose associated with the dialyzable glycopeptides was found in this case (Table IV).The increased level of hexose was due to an elevation of mannose ; galactose levels remained normal. The results suggest that in Tay-Sachs disease, the catabolism of the non-dialyzable glycopeptides ceases after the galactose residues had been removed. The resultant mannose-rich glycopeptides accumulate and appear in the dialyzable glycopeptide fraction.

Finally, it might be mentioned that in metachromatic leukodystrophy (deficiency of lysosomal sulfatase), one would expect an increase in glycopeptide chains which contain sulfate-ester groups.

REFERENCES

1 ADEY, W.R., in Biocybernetics on the Central Nervous System (Ed. PROCTOR, L.D.), Little-Brown and Co., Boston, 1969, p.1
2 ARIMA, T., MURAMATSU, T., SAIGO, K. and EGAMI, F., Jap. J. Exp. Med., 39 (1969) 301.
3 ARSENI, C. and CARP, N., Rev. Roum. Neurol., 6 (1969) 223.
4 BARONDES, S.H., J. Neurochem., 15 (1968) 699.
5 BARONDES, S.H. and DUTTON, G.R., J. Neurobiol., 1 (1969) 99.
6 BOGOCH, S., in The Biochemistry of Memory, Oxford Univ. Press, New York, 1968.
7 BOSMANN, G.B., Life Sciences, 9 (1970) 851.
8 BOSMANN, H.B., CASE, K.R. and SHEA, M.B., Febs Letters, 11 (1970) 261.
9 BOSMANN, H.B., HAGOPIAN, A. and EYLAR, H., Arch. Biochem. Biophys., 130 (1969) 573.
10 BOSMANN, H.B. and HEMSWORTH, B.A., J. Biol. Chem., 245 (1970) 363.
11 BOSMANN, H.B. and MARTIN, S.S., Science, 164 (1967) 190.
12 BRUNNGRABER, E.G., in Handbook of Neurochemistry (Ed. LAJTHA,A.) Plenum Press, New York, 1969, vol. 1, p.223.
13 BRUNNGRABER, E.G., Perspect. Biol. Med., 12 (1969) 467.
14 BRUNNGRABER, E.G., in Protein Metabolism of the Nervous System (Ed. LAJTHA, A.) Plenum Press, New York, 1970, p.383.
15 BRUNNGRABER, E.G., J. Pediatrics, 77 (1970) 166.
16 BRUNNGRABER, E.G., in Biogenesis, Evolution and Homeostasis (Ed. LOCKER, A.) Springer Verlag, Heidelberg, in press.
17 BRUNNGRABER, E.G., AGUILAR, V. and ARO, A., Arch. Biochem. Biophys., 129 (1969) 131.
18 BRUNNGRABER, E.G., ARO, A. and BROWN, B.D., Clin. Chim. Acta, 29 (1970) 233.
19 BRUNNGRABER, E.G. and BROWN, B.D., Biochim. Biophys. Acta, 69 (1963) 581.
20 BRUNNGRABER, E.G. and BROWN, B.D., J. Neurochem., 11 (1964) 449.
21 BRUNNGRABER, E.G. and BROWN, B.D., Biochim. Biophys. Acta, 83 (1964) 357.
22 BRUNNGRABER, E.G. and BROWN, B.D., Biochem. J., 103 (1967) 65.
23 BRUNNGRABER, E.G., BROWN, B.D. and AGUILAR, V., J. Neurochem., 16 (1969) 1059.
24 BRUNNGRABER, E.G., BROWN, B.D. and CHANG, I., J. Neuropath. Exptl. Neurol., 30 (1971) 525.
25 BRUNNGRABER, E.G., BROWN, B.D. and HOF, H., Clin. Chim. Acta, 32 (1971) 159.
26 BRUNNGRABER, E.G., DEKIRMENJIAN, H. and BROWN, B.D., Biochem. J., 103 (1967) 73.
27 BRUNNGRABER, E.G. and WHITNEY, G., J. Chromatog., 32 (1968) 749.
28 BRUNNGRABER, E.G., WITTING, L.A., HABERLAND, C. and BROWN,B.D., Brain. Res., 38 (1972) 151.
29 CASTEJON, H.V., Acta Histochem., 38 (1970) 55.

30 DEN, H., KAUFMAN, B. and ROSEMAN, S., J. Biol. Chem., 245 (1971) 6607.
31 DEKIRMENJIAN, H. and BRUNNGRABER, E.G., Biochim. Biophys. Acta, 177 (1969) 1.
32 DEKIRMENJIAN, H., BRUNNGRABER, E.G., JOHNSTON, N.L. and LARRAMENDI, L.M.H., Exptl. Brain Res., 8 (1969) 97.
33 DEKIRMENJIAN, H. and LIU, T., Feder. Proc., 30, N°3,part II (1971) 1194 Abstr.
34 DE VRIES, G.H. and BARONDES, S.H., J. Neurochem., 18 (1971) 101.
35 DI BENEDETTA, C., BRUNNGRABER, E.G., WHITNEY, G., BROWN, B.D. and ARO, A., Arch. Biochem. Biophys. 131 (1969) 404.
36 DI BENEDETTA, C., CHANG, L. and BRUNNGRABER, E.G., Ital. J. Biochem., in press.
37 DI BENEDETTA, C., IMBIMBO, B., CAPUTO, U. and SANTAMARIA, R., Russ. Med. Sper., Suppl. 15 (1968) 35.
38 DISCHE, Z.,in Protides of the Biological Fluids (Ed.PEETERS,H.), Elsevier Publ. Co., Amsterdam, 1966, vol.13, p.1.
39 DISCHE, Z., KAHN, N., ROTHSCHILD, C., DANILCHENKO, A., LICKING, J., and CHANG, S.C., J. Neurochem., 17 (1970) 649.
40 DROZ, B. in Protein Metabolism of the Nervous System (Ed. LAJTHA, A.) Plenum Press, New York, 1970, p.93.
41 DUNN, J.T. and SPIRO, R.G., J. Biol. Chem., 242 (1967) 5556.
42 DURAND, P., BORRONE, C. et DELLA CELLA, G., J. Pediatrics, 75 (1969) 665.
43 DUTTON, G.R. and BARONDES, S.H., J. Neurochem., 17 (1970) 913.
44 ELAM, J.S., GOLDBERG, J.M., RADIN, N.S. and AGRANOFF, B.W., Science, 170 (1970) 458.
45 FAMBROUGH, D.M., HARTZELL, C. and MUSSELMAN, A., Carnegie Instit. Washington, Yearbook, 1968-69, p.531.
46 FISHER, S., CELLINO, M., ZAMBRANO, Z., ZAMPIGHI, G., TELLEZ-NAGEL, M., MARCUS, D. and CANESSA-FISHER, M., Arch. Biochem. Biophys., 138 (1970) 1.
47 FORMAN, D.S., McEWEN, B.S. and GRAFSTEIN, D., Brain Res., 28 (1971) 119.
48 GLASSMAN, E., Ann. Rev. Biochem., 38 (1969) 605.
49 GOUSSAULT, Y. and BOURRILLON, R., Biochem. Biophys. Res. Commun., 40 (1970) 1404.
50 GROSSFELD, R.M. and SHOOTER, E.M., J. Neurochem.,18 (1971) 2265.
51 HABERLAND, C. and BRUNNGRABER, E.G., Arch. Neurol., 23 (1970) 481.
52 HABERLAND, C., BRUNNGRABER, E.G., WITTING, L.A., and HOF, H., Neurology, in press.
53 HABERLAND, C., PEROU, M., BRUNNGRABER, E.G. and HOF, H., J. Neuropathol. Exptl. Neurol., 30 (1971) 431.
54 HATCHER, V.B. and McPHERSON, C.F.C., J. Immunol., 104 (1970) 633.
55 HELLER, J., Biochemistry, 8 (1969) 675.
56 HIRANO, S., HOFFMAN, P. and MEYER, K., J. Org. Chem., 26 (1961) 5064.

57 HOLIAN, O., DILL, D. and BRUNNGRABER, E.G., Arch. Biochem. Biophys., 142 (1970) 111.
58 HOLIAN, O., ROUTTENBERG, A. and BRUNNGRABER, E.G., Life Sci. 10 (1971) 1029.
59 IRWIN, L.N. and SAMSON, F.S., J. Neurochem., 18 (1971) 203.
60 JAMIESON, G.A., J. Biol. Chem., 240 (1965) 2914.
61 KASHKIN, K.P., SHARETSKII, A.N., SURIMOV, B.P. and BOCHKOVA, D.N., Vop. Med. Khim., 15 (1969) 235.
62 KASHNIG, D.M. and KASPAR, C.B., J. Biol. Chem., 244 (1969) 3786.
63 KOENIG, H. in The Lysosomes in Biology and Pathology (Ed. DINGLE, J.T. and FELL, H.B.) North-Holland, Amsterdam, 1969, vol.2, p.110.
64 KOGAN, A.B., SHIBKOVA, S.A., ALEINIKOVA, T.V. and KHUSAINOVA, I.S., Funkts. Neurokhim. Tsent.Nerv. Sist., Mater. Vses. Simp. 1st, 1966, p.102.
65 LANGLEY, O.K., J. Neurochem., 17 (1970) 1535.
66 LIAKOPOULOU, A. and MacPHERSON, C.F.C., J. Immunol. 105 (1970) 512.
67 MAIZEL, J.V., Science, 151 (1966) 988.
68 MARGOLIS, R.K. and MARGOLIS, R.U., Biochemistry, 9 (1970) 4389.
69 MATTHEWS, M.B. and CIFONELLI, J.A., J. Biol. Chem., 240 (1965) 4140.
70 McEWEN, B.S. and HYDEN, H., J. Neurochem., 13 (1966) 823.
71 MIHAILOVIC, M.L., DIVAC, I., MITROVIC, K., MILOSEVIC, D. and JANKOVIC, B.D., Exptl. Neurol., 24 (1969) 325.
72 OCKERMAN, P.A., J. Pediatrics, 75 (1969) 360.
73 OCKERMAN, P.A., J. Pediatrics, 77 (1970) 168.
74 PASQUALINO, A. and TESSITORE, V., Biol. Lat., 21 (1968) 1.
75 PELLEGRINO DI IRALDI, A. and RODRIGUEZ DE LORES ARNAIZ, G., J. Neurochem., 17 (1970) 1601.
76 QUARLES, R.H. and BRADY, R.C., J. Neurochem., 17 (1970) 801.
77 RAJAM, P.C., GAUDREAU, C.J., GRADY, A. and RUNDLETT, St., Immunology, 17 (1969) 813.
78 RAMBOURG, A., HERNANDEZ, W. and LEBLOND, C.P., J. Cell Biol., 40 (1969) 395.
79 RICHMOND, J.E., GLASER, R.M. and TODD, P., Exptl. Cell Res., 52 (1968) 43.
80 RIDDLE, D. and LEONARD, B.E., Neuropharmacol., 9 (1970) 283.
81 ROUKEMA, P.A. and HEIJLMAN, J., J. Neurochem., 17 (1970) 773.
82 ROUKEMA, P.A., VAN DEN EIJNDEN, D.H., HEIJLMAN, J. and VAN DEN BERG, G., FEBS Letters, 9 (1970) 267.
83 ROUTTENBERG, A., ZECHMEISTER, E.B. and BENTON, C., Life Sci., 9 (1970) 909.
84 ROUX, M., C.R. Acad. Sci.,Sér.D., 270 (1970) 717.
85 SARASWATHI, S. and BACCHAWAT, B.K., Biochim. Biophys. Acta, 212 (1970) 170.
86 SCHACHTER, H., JABBAL, I., HUDGIN, R.L., PINTERIC, L.,McGUIRE, E.J. and ROSEMAN, S., J. Biol. Chem., 245 (1970) 1090.
87 SEIJO, L. and RODRIGUEZ DE LORES ARNAIZ, G., Biochim. Biophys. Acta, 211 (1970) 595.

88 SELLINGER, O.Z. and NORDRUM, L.M., J. Neurochem., 16 (1969) 1219.
89 SHAPIRO, A.L., VINUELA, E. and MAIZEL, J.V., Biochem. Biophys. Res. Commun., 28 (1967) 814.
90 SHULMAN, S., MILGROM, F. and SWANBORG, R.H., Immunology, 16 (1969) 25.
91 SPIRO, R.G., J. Biol. Chem., 237 (1962) 646.
92 SPIRO, R.G., J. Biol. Chem., 239 (1964) 567.
93 SPIRO, R.G., J. Biol. Chem., 240 (1965) 1603.
94 SPIRO, R.G., Ann. Rev. Biochem., 39 (1970) 599.
95 STERBA, G. and WOLF, G., Histochemie, 17 (1969) 57.
96 SUSZ, J., Feder. Proc., 28 (1969) 830.
97 SUZUKI, K., J. Neurochem., 12 (1965) 629.
98 SWANBORG, R.H. and SHULMAN, S., Immunology, 19 (1970) 31.
99 VACCARI, A. and CUGARRA, F., Biochem. Pharm., 17 (1968) 2399.
100 VACCARI, A. and VERTUA, R., Biochem. Pharm., 19 (1970) 2105.
101 VAN HOOF, F. and HERS, H.G., Europ. J. Biochem., 7 (1968) 34.
102 WAGH, P.V., BORNSTEIN, I. and WINZLER, R.J., J. Biol. Chem., 244 (1969) 658.
103 WARECKA, K., J. Neurochem., 17 (1970) 829.
104 WARECKA, K. and HANZAL, F., Cesk. Neurol., 32 (1969) 25.
105 WHETSELL Jr., W.O. and BUNGE, R.P., J. Cell Biol., 42 (1969) 490.
106 WILSON, L., BRYAN, J., RUBY, A. and MAZIA, D., Proc. Natl.Acad. Sci. US., 66 (1970) 807.
107 YASUI, K., NOZIMA, T., HOMMA, R. and VEDA, S., Acta Virol. (Prague)(Engl. Ed.), 13 (1969) 158.
108 ZATZ, M. and BARONDES, S.H., J. Neurochem., 17 (1970) 157.

GLYCOPROTEINS OF THE SYNAPTIC REGION

W.C. BRECKENRIDGE, J.E. BRECKENRIDGE and I.G. MORGAN

Centre de Neurochimie du CNRS, 67-Strasbourg (France)

Glycoproteins have been found in most tissues as components of soluble and membranous fractions of the cell. Vast quantities of PAS-staining material have been demonstrated by electron microscopy (35) in cell membranes and in particular, the plasma membrane appears to be enriched in protein-bound carbohydrate. The carbohydrate composition of plasma membranes from various tissues has been reported (6, 20, 21, 24, 29, 33, 37). Brain glycoproteins have already been extensively studied (for review, see ref. 13) and evidence has been presented indicating that subcellular fractions enriched in neuronal plasma membrane are also enriched in protein-bound carbohydrate (17). However no detailed studies of glycoproteins and glycopeptides have been carried out on homogeneous fractions from brain.

In general, glycoproteins of plasma membranes have been implicated in contact inhibition (1) and cell differentiation (2). It has also been suggested (3, 7) that the glycoproteins of neuronal plasma membranes are important in learning and memory ; functions which could depend on some form of interaction between neuronal surfaces. If these phenomena do result from subtle molecular interactions between neuronal surfaces, glycoproteins and possibly glycolipids, as shown by their importance in other systems (24, 25, 30) would be ideal candidates for mediating the processes. Studies on the roles played by possible close interactions of glycoproteins of neuronal plasma membranes in the complex processes of neuronal maturaton and function obviously depend upon an exact knowledge of the structure and arrangement of the neuronal plasma membrane glycoproteins and their carbohydrate chains.

Research on glycoproteins of neuronal plasma membranes has been hindered by the problem of the preparation of this subcellular fraction in a sufficiently pure form. The present communication describes initial studies on the characterization of the polysaccharide chains of glycoproteins present in synaptosomal plasma membrane, which we are using as an experimental system to study the account possible specializations of the synaptosomal plasma membrane, in particular in the region of the synaptic cleft.

Preparation of Synaptosomal Subfractions

A detailed description of the preparation and characterization of the synaptosomal plasma membrane has been published elsewhere (34). The procedure is similar to the method described by Whittaker, Michaelson and Kirkland (42) but has been modified in certain respects. The crude mitochondrial pellet was washed three times to remove extensive microsomal contamination. Synaptosomes were isolated on isotonic Ficoll gradients instead of sucrose density gradients, since such synaptosomes are more fragile to osmotic shock. Following osmotic shock, the membranous constituents of the synaptosomes were pelleted and then placed on sucrose density gradients. The various membrane fractions harvested from the gradient, were pelleted, resuspended in dilute buffer and then centrifuged under approximately the same conditions as were used to spin down the crude mitochondrial pellet (Fig. 1). This removed all material whose weight had not been changed by osmotic shock. Finally the membranes were sedimented and then washed 5 times to remove contaminating sucrose.

Fractions F and G were greatly enriched in the putative plasma membrane markers ; (Na^+, K^+)-ATPase and gangliosides (Table I).

TABLE I

Enrichment of Plasma Membrane Markers

Marker	Fraction[1]		
	F	G	Homogenate
Na,K-dependent ATPase[2]	103.1	63.5	8.90
Ganglioside[3]	39.8	24.0	3.60

[1]Fractions F and G represent membranes banding at the interphase between 0.6-0.8 and 0.8-1.0 M sucrose respectively. [2]μmole substrate consumed/mg protein/hour. [3]μg lipid sialic acid/mg protein.

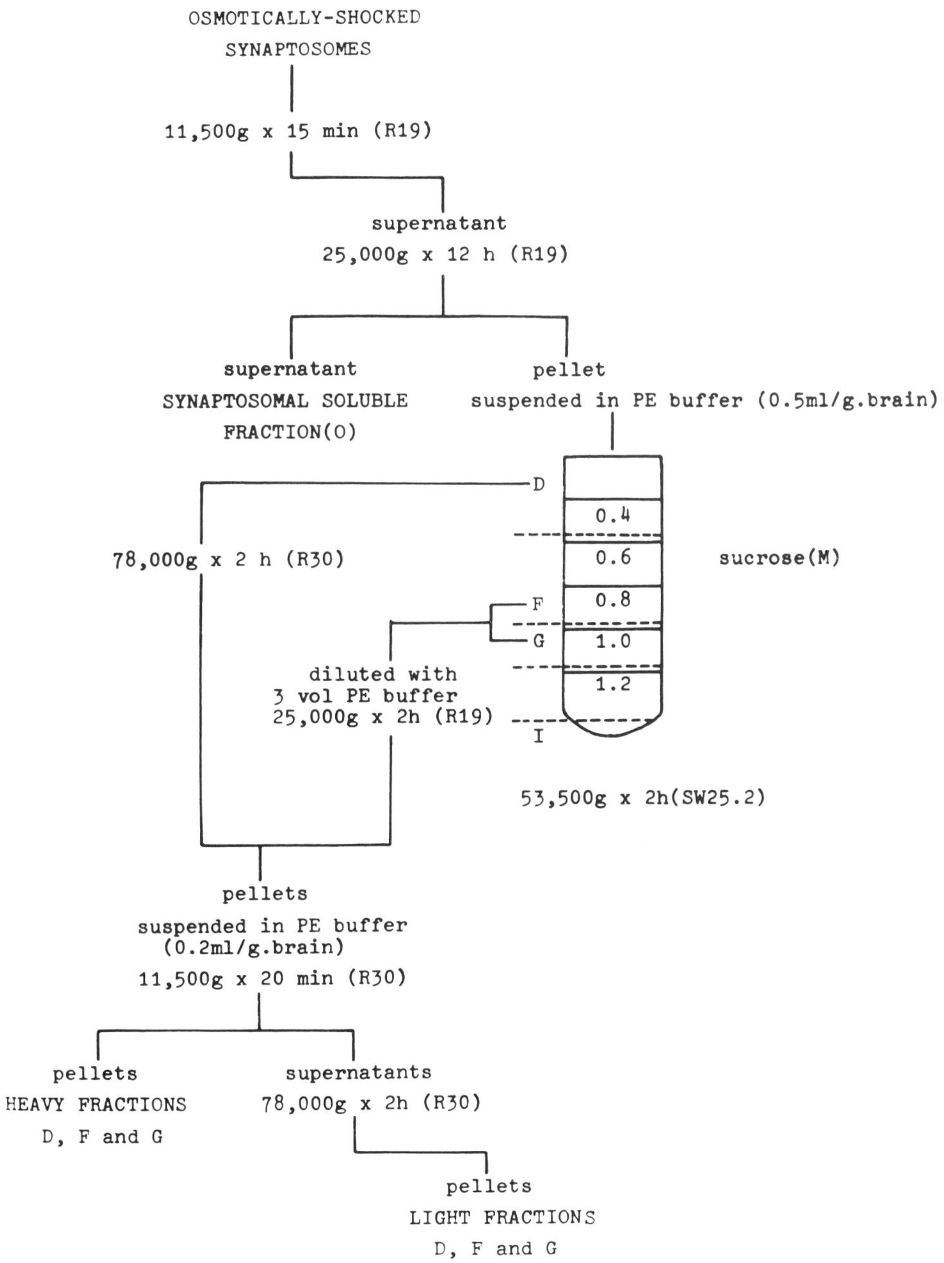

Fig. 1. Scheme of preparation of subsynaptosomal fractions. PE buffer is 0.1 mM EDTA in 1 mM potassium phosphate (pH 7.6).

It was found that fraction I (the pellet) was enriched in mitochondria. Fraction D was enriched in synaptic vesicles but still contained some plasma membrane (34). On the basis of many enzymatic and chemical markers (Table II) fraction F was estimated to be over 80 percent synaptosomal plasma membrane. There was no significant contamination with rough endoplasmic reticulum, soluble proteins, lysosomes or myelin. In addition, no significant amounts of Golgi apparatus were found (unpublished observations). Significant contaminants were outer mitochondrial membrane (< 5 %), smooth endoplasmic reticulum (< 5 %) and glial plasma membrane (< 10 %). Electron microscopy (Fig. 2) of fraction F showed a homogeneous population of smooth surfaced membraneous vesicles. In addition to analyses of these highly purified fractions, the material which was sedimented at 11,500 X g for 20 min from each fraction (Fig. 1) was also analyzed. While not as rich as F and G "light" fractions, the "heavy" fractions corresponding to F and G were 6-8-fold enriched in ganglioside and (Na^+,K)-ATPase. The most significant contaminant in the heavy fractions appeared to be myelin. As discussed later this material was also used for the characterization of glycopeptides.

TABLE II

Assessment of Contamination of Synaptosomal Plasma Membranes†

Contaminant	Percent contamination	Marker used
Inner mitochondrial membrane	<1	-Cytochrome c oxidase Succinate dehydrogenase
Outer mitochondrial membrane	<5	-Monoamine oxidase Rotenone insensitive NADH Cytochrome c reductase
Microsomes	<5	-RNA Rotenone insensitive NADPH: cytochrome c reductase
Soluble proteins	<1	-Lactate dehydrogenase
Lysosomes	<1	-β-glucosidase β-galactosidase
Myelin	<1	-Cerebrosides and sulphatides
Glial membranes	<10	-2',3'-cyclic AMP-3'-phosphohydrolase S-100 protein Long chain fatty acids of sphingomyelin

†Results are based on analyses of fractions F and G.

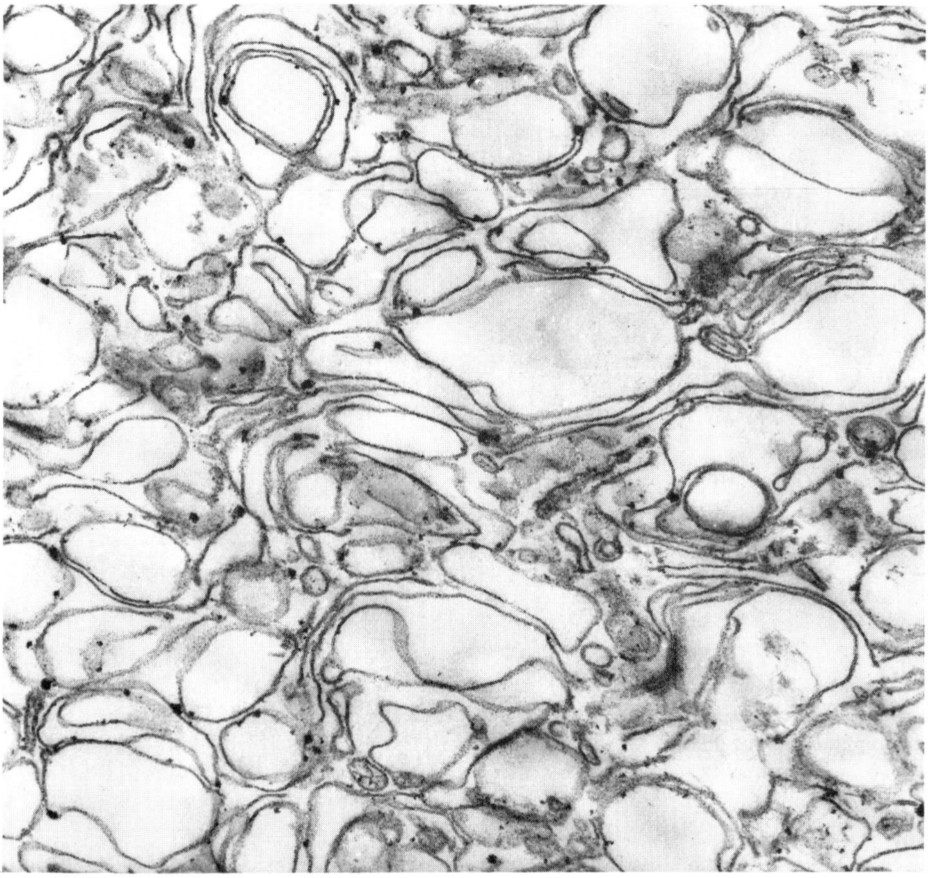

Fig. 2. Typical field of fraction F. X 30,000.

Protein-Bound Carbohydrate Composition of Synaptosomal Subfractions

After extraction of the lipids, the carbohydrate composition of the protein residues was determined. For small quantities of material, the protein was recovered by filtering the chloroform-methanol extract through a Millipore teflon disc (43). The content of protein-bound sialic acid in various synaptosomal subfractions is shown in Table III. Fraction F contained the greatest amount of sialic acid per mg protein, while fractions E and G were somewhat less rich. This distribution of sialic acid paralleled the distribution of (Na^+,K^+)-ATPase and gangliosides. Protein sialic acid was enriched approximately 5-7-fold in the plasma membranes when compared to total brain homogenate (15, 17). The level of protein-bound sialic acid in the synaptosomal plasma membranes was similar to the levels in plasma membranes which have been isolated from

TABLE III

Protein-Bound Sialic Acid of Synaptosomal Subfractions

Subfraction[1]		Sialic acid[2]
Soluble protein	O	0.3
Vesicles	D	19.0
Plasma membranes	F	41.1
	G	27.8
Mitochondria	I	2.0

[1]Fractions defined in figure 1.
[2]After extraction of the lipids the protein residues were hydrolyzed with 0.1 N H_2SO_4 at 80° for 1 h. Sialic acid was estimated by the method of Warren (42). Results are expressed as nmole sialic acid/mg protein. The recovery of sialic acid was 95 %.

other cells (21, 24, 29, 33). Fraction D, which was enriched in synaptic vesicles, also contained some protein-bound sialic acid. Some, but not all of this may have been due to contamination of the vesicle fraction with plasma membranes. The mitochondrial fraction contained the lowest levels of sialic acid of the membrane fractions. Recent reports (16) suggest that some protein-bound sialic acid is found in the inner and outer membranes of liver mitochondria but, in our experiments, contamination with plasma membranes cannot be excluded. The 5 percent contamination of fraction F with outer mitochondrial membranes probably does not contribute much sialic acid to this fraction since the sialic acid content of outer mitochondrial membranes from rat liver (16) is very much lower than that of fraction F. The synaptosomal soluble fraction contained the lowest amount of sialic acid ; only 1 percent of the level in fraction F.

In view of the fact that fractions F and G were the purest fractions of plasma membrane, further carbohydrate analyses were performed only on them. After appropriate hydrolytic procedures, the neutral and amino sugars were analyzed by chromatography on polycarbonate sheets (32). The results (Table IV) again showed that the highest concentration of all sugars, with the exception of mannose, was found in fraction F. It has been suggested that brain glycoproteins might contain mannosamine but this sugar could not be unequivocally identified due to the lack of an adequate method (11). In the present system standard mannosamine was readily separated from glucosamine. Analyses of both the membranes and glycopeptides indicated that mannosamine was not a component of these

TABLE IV

Protein-Bound Carbohydrate Composition of Rat Brain Plasma Membranes

Carbohydrate[1]	Fraction[2]			
	F		G	
	nmole/mg protein	Molar ratio	nmole/mg protein	Molar ratio
Sialic acid	41.1	1.00	27.8	1.00
Fucose	21.4	0.52	15.2	0.55
Glucosamine	128.0	3.10	89.0	3.20
Galactosamine	34.1	0.83	13.0	0.47
Mannose	40.6	0.99	46.1	1.66
Galactose	33.3	0.81	25.5	0.92

[1]Carbohydrates were estimated after appropriate hydrolytic conditions (9). Hexosamines and hexoses were estimated by optical densitometry after chromatography on polycarbonate (32).
[2]Fractions are defined in table I.

glycoproteins. The quantity of mannose exceeded the amount of galactose. Considerable quantities of glucose were also found. However subsequent experiments on the glycopeptides indicated the presence of a high molecular weight glucose polymer. It is probable that this polymer is either glycogen or Ficoll and thus no values have been included for glucose. Glucosamine was the major hexosamine, indicating that gangliosides were completely extracted. The total protein-bound carbohydrate, by weight, was 60-80 μg/mg protein.

The molar ratios of the carbohydrate components show that glucosamine is present in the highest quantity followed by sialic acid and the neutral sugars. These ratios can be compared to preparations from total rat (18), bovine (17) and human (27) brains. In all cases the carbohydrate ratios of the total brain were not identical to those of the synaptosomal plasma membranes. In particular the content of hexoses was lower in the synaptosomal preparations. Such differences would be expected since total brain preparations contain carbohydrate from all subcellular fractions of both neuronal and glial cells. The plasma membranes from liver (21) contain higher ratios of hexoses and lower ratios of hexosamines while plasma membranes from rat ascites tumor contain higher levels of hexoses (37).

Analysis of Glycopeptides

There are essentially two approaches to the study of the structure of glycoproteins. One could attempt to isolate the individual glycoproteins. However purification of membrane proteins is difficult since the proteins are insoluble in mild aqueous environments. Although they can be solubilized by detergents, the latter interfere with many of the conventional methods involved in the fractionation of mixtures of proteins. However it is possible to study the carbohydrate portion of the glycoproteins by first digesting the membrane proteins with pronase, or some other proteolytic enzyme, and subsequently isolating the soluble glycopeptides (Fig. 3.). The lipid-free membranes were digested with pronase which solubilized virtually all the carbohydrate material (8,9). Nucleotides and mucopolysaccharides were eliminated by precipitation with cetyl pyridinium chloride and the excess reagent was removed by extraction with pentanol (36). When the aqueous phase was passed through a column of Sephadex G-15, virtually all of the hexose and sialic acid was eluted in the void volume while lower molecular weight peptides, amino acids and traces of sucrose, if present, were eluted in the second and third void volumes.

This crude glycopeptide fraction was placed on a column of Sephadex G-50 and eluted with an ammonium formate buffer(Fig. 4.). The hexose pattern and sialic acid pattern showed two characteristic groups of glycopeptides. This pattern was found for both light and heavy fractions of F and G. There was always a hexose positive peak in the void volume. However this material contained only glucose. It was assumed that this peak represented either glycogen or possibly Ficoll which had been bound to the membranes. Subsequently there were two broad peaks of hexose positive material with different molecular weights. Virtually all of the sialic acid was confined to the higher molecular weight group while the second peak could be isolated almost free of sialic acid. Due to the incomplete resolution of these two classes the carbohydrate ratios of these two peaks, which were very different, could be influenced by the exact positions at which the fractions were collected. The small peak at the end of the elution was probably due to small amounts of residual sucrose. The absorption at 230 nm showed that the majority of the peptide material was associated with low molecular weight compounds.

The column fractions were pooled into two fractions (S-I and S-II) and analyzed for their carbohydrate content. Peak S-I was extremely rich in sialic acid (Table V) while the relative proportions of hexoses and hexosamines were lower than those obtained by analysis of the total membranes (Table VI). By contrast peak S-II contained glucosamine and mannose as the major sugars. The first peak contained slightly more of the total hexose (60 %) than peak S-II (40 %). Essentially all of the fucose was present in peak S-I.

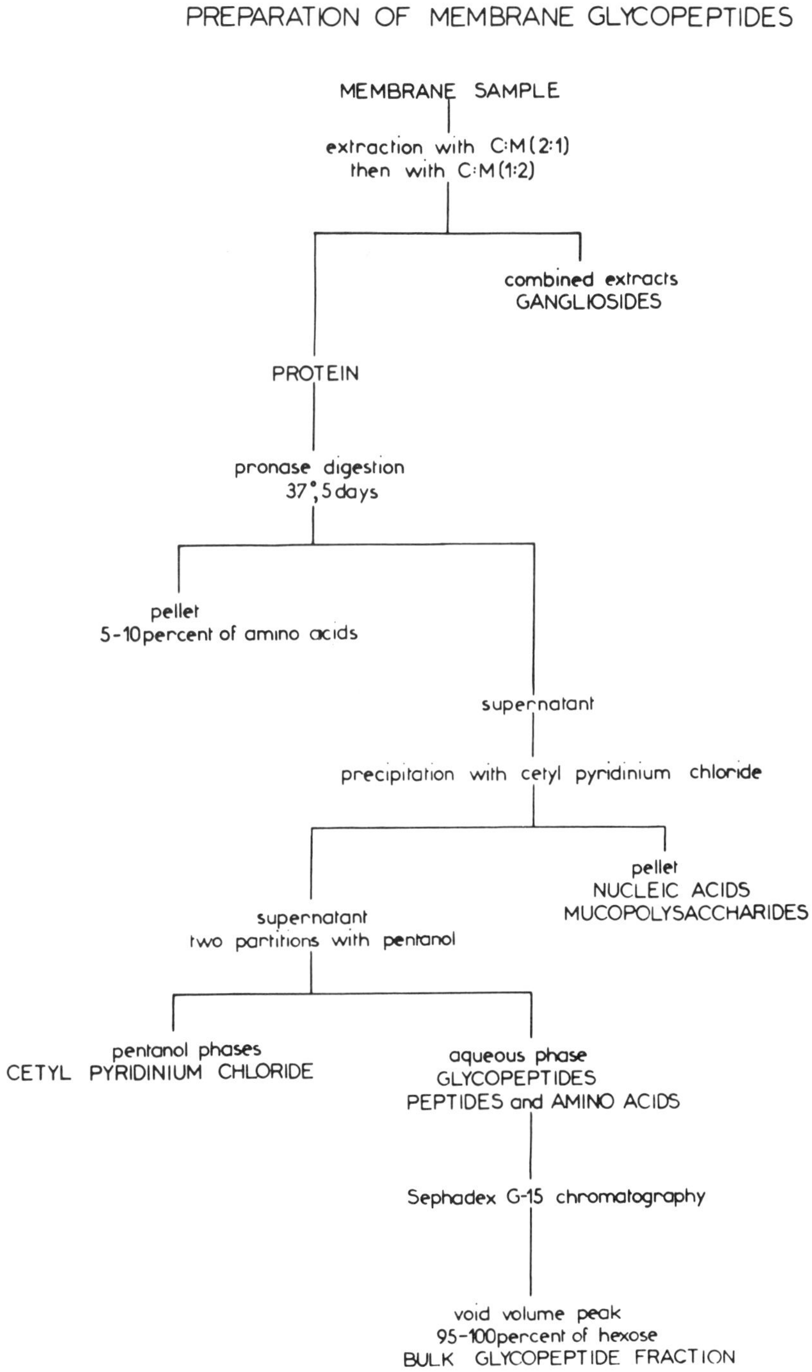

Fig. 3. Preparation of crude glycopeptides. C- chloroform ; M-methanol.

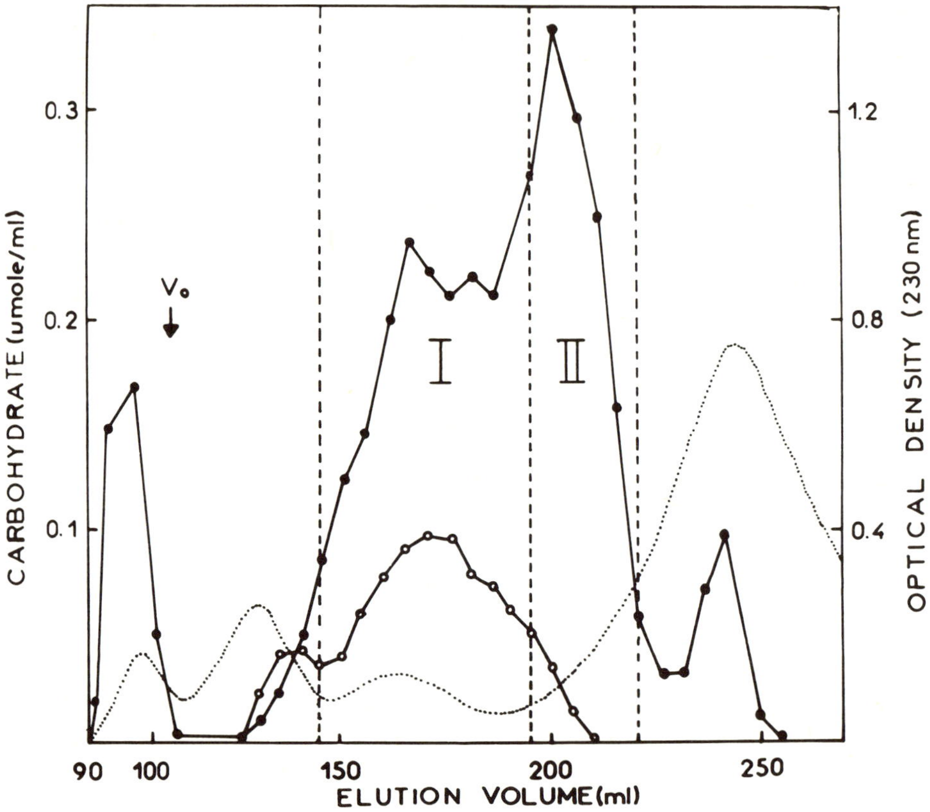

Fig. 4. Chromatography on Sephadex G-50 of total glycopeptides from synaptosomal plasma membranes. oooo hexose ; oooo sialic acid ; .... absorption 230 nm.

The sum of the carbohydrates of peak S-I and S-II gave a composition essentially the same as that of the total membranes.

The compositions of the glycopeptides are, in some respects, similar to those reported by Di Benedetta et al. (18) for the glycopeptides of total rat brain. In agreement with these studies, the higher molecular weight glycopeptides contain proportionally more sialic acid while the low molecular weight species are rich in hexosamine and mannose. However the carbohydrate ratios of the synaptosomal plasma membrane glycopeptides are distinctly different from fractions prepared from total rat brain. In general the amounts of hexose and hexosamine are lower relative to sialic acid in peak I while the lower molecular weight glycopeptides contain much higher levels of glucosamine relative to mannose than those isolated from total brain.

TABLE V

Carbohydrate Composition of Synaptosomal Plasma Membrane Glycopeptides[1]

Carbohydrate	Glycopeptide Fraction	
	I	II
Sialic acid	1.00	1.00
Fucose	0.50	< 0.10
Glucosamine	2.00	11.20
Galactosamine	0.51	1.20
Mannose	0.72	8.00
Galactose	0.51	1.30
Glucose	0.10	0.50

[1]Results expressed as molar ratios relative to sialic acid.

TABLE VI

Carbohydrate Composition after Resolution on DEAE-Sephadex of Hexose-Hexosamine Rich Glycopeptides from Synaptosomal Plasma Membranes[1]

Carbohydrate	Fraction[2]		
	SIIa	SIIb	SIIc
Glucosamine	1.00	1.00	1.00
Galactosamine	0.10	0.07	0.05
Mannose	1.79	1.32	2.66
Galactose	0.06	0.15	0.17
Glucose	0.06	0.14	0.27

[1]Results expressed as molar ratios relative to glucosamine.
[2]Fractions were collected as described in Fig. 5.

In both peaks the major hexosamine is glucosamine. However Di Benedetta *et al*. (18) have claimed that the low molecular weight glycopeptides from total brain have galactosamine as the major hexosamine. In parallel studies to these (9) it has been found that the microsomal fractions contain low molecular weight glycopeptides with a glucosamine mannose ratio of 1:2:5. Galactosamine was not the major hexosamine. This once again indicates that the different subcellular fractions may contain glycoproteins with different carbohydrate compositions.

Analyses of the Mannose, Glucosamine-Rich Glycopeptides

Analyses (8) of the sialic acid-rich glycopeptides from microsomes using ion exchange chromatography showed them to be a complex mixture of molecular species containing different amounts of fucose and sialic acid. However in view of the complexity of the fraction, contrasted to the simplicity of the peak II glycopeptides (Fig.5.),

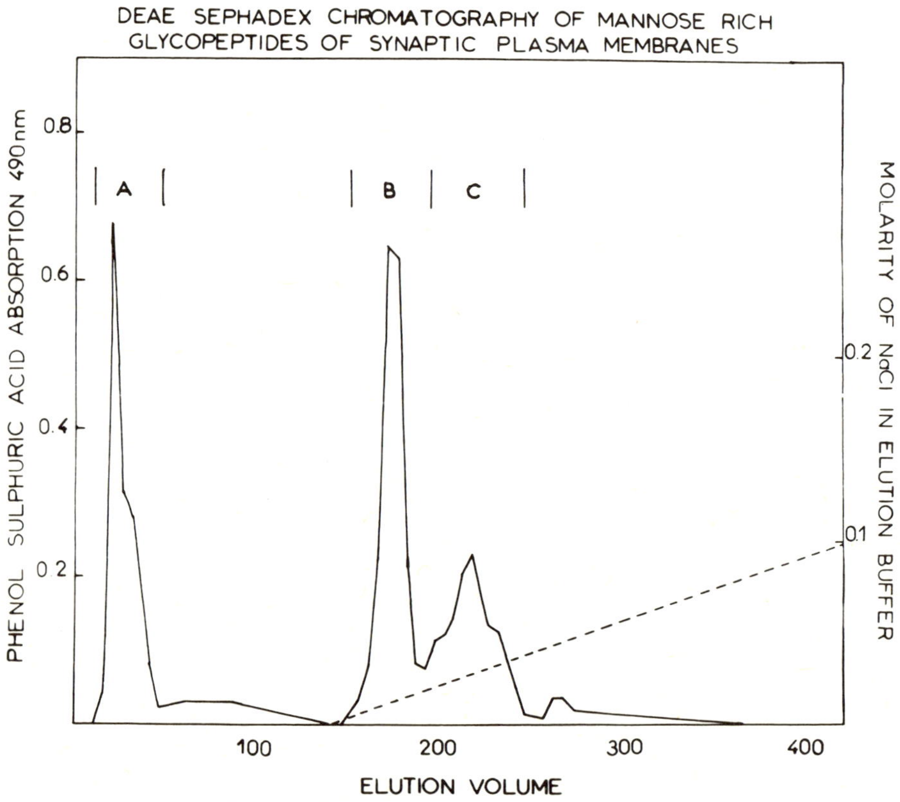

Fig. 5. Chromatography on DEAE-Sephadex A-25 of S-II glycopeptides. ——— absorption of phenol-sulphuric acid reaction at 490 nm ; ------ sodium chloride gradient.

we have concentrated on studying the latter fraction. After hydrolyzing the S-II glycopeptides from the "heavy" fractions of F and G with leucine amino peptidase and carboxypeptidase the S-II glycopeptides were placed on a column of DEAE-Sephadex A-25. One peak was eluted with the starting buffer (0.005 M Tris HCl, pH 8.0). Further elution with a salt gradient gave two additional peaks of carbohydrate material along with several very minor peaks (Fig. 5.). The resolution of mannose-rich glycopeptides (peak II) from microsomes also gave the same three peaks (9). Analyses of the three major peaks derived from the synaptosomal plasma membrane (Table VI) and the microsomal fractions, (Table VII) showed that there were differences in carbohydrate composition between the three peaks from a given fraction and between the corresponding peaks from the two subcellular fractions. In the plasma membrane fraction the ratio of mannose : glucosamine was slightly greater than 1 in peak A (1.79) and B (1.32). Whole peak C contained considerably more mannose than glucosamine (2.66 : 1). In contrast peaks A and B from the microsomal glycopeptides (Table VII) had ratios of mannose:glucosamine of 4:1 and 3:1 while peak C had a similar ratio to that found for peak C of the

TABLE VII

Carbohydrate Composition after Resolution on DEAE-Sephadex of Hexose-Hexosamine Rich Glycopeptides from Microsomes[1]

Carbohydrate	Fraction[2]		
	MIIa	MIIb	MIIc
Glucosamine	1.00	1.00	1.00
Galactosamine	0.04	0.16	0.04
Mannose	3.99	2.86	2.66
Galactose	0.09	0.25	0.49
Glucose	0.09	0.25	0.49
Sialic acid	0.00	0.00	0.12

[1]Results expressed as molar ratios relative to glucosamine.
[2]Fractions were collected as described in Fig.5.

plasma membrane fraction. None of the peaks contained sialic acid or fucose, and only small amounts of galactose and glucose were found. Considerable problems arise in estimating glucose since all of the resolution procedures use glucose-type polymers. It is possible that small amounts of glucose are eluted from the columns. After extensive purification, small amounts of glucose were still present. Further studies will be required to determine whether glucose is a true constituent of the synaptosomal plasma membrane glycopeptides but it is not a major component of any glycopeptide fraction.

The above results indicate definite differences in the structure of glycopeptides in the microsomal fraction and the synaptosomal plasma membrane fraction. The latter contain considerably more glucosamine in relation to mannose and peaks A and B represent a larger proportion of the total mannose-rich glycopeptides than in the microsomal fraction. Only peak C appears to have a similar sugar ratio in both subcellular fractions. It is probable that the microsomal fraction contains some synaptosomal and neuronal plasma membrane fragments. It is also possible that all the glycopeptide fractions which have been isolated from DEAE-Sephadex are still mixtures containing carbohydrate chains of varying length and varying sugar ratios. Low molecular weight glycopeptides, similar to those present in the synaptosomal plasma membrane, have been isolated from total rat brain by Brunngraber _et al_. (11). It was reported that they contained 8 mole of hexose (mainly mannose) and 4 mole of hexosamine per glycopeptide molecule. The low molecular weight glycopeptides from the microsomes appear to contain species with higher ratios of mannose to glucosamine. This may be due to the presence of glycoproteins, in the smooth endoplasmic reticulum, with only partially completed carbohydrate chains.

The present results are in contrast to recent analyses of liver plasma membranes (31), which indicated that only glycopeptides containing sialic acid and fucose were present. It is possible that the glycopeptides, which contain only glucosamine and mannose, are characteristic of the neuronal plasma membrane, however more data will be required to establish such a possibility.

Selective Extraction of Membrane Glycoproteins

In soluble glycoproteins the carbohydrate chains exhibit microheterogeneity (5,19, 26, 39). Thus the carbohydrate chains at a given point in a polypeptide chain may vary in carbohydrate composition. In many cases a "completed carbohydrate chain" has been shown to contain fucose or sialic acid as terminal sugars preceeded by galactose and N-acetylglucosamine and then an internal region containing mannose and N-acetylglucosamine or N-acetylgalactosamine (39). Usually glycoproteins from the plasma or other sources contain a small portion of their carbohydrate chains uncompleted with only the internal region of N-acetylglucosamine and mannose. This phenomenon has been attributed to the inability of the glycosyl transferases to complete all the carbohydrate chains in a given protein during the passage of the proteins through the smooth endoplasmic reticulum and the Golgi apparatus (39). More recently several glycopeptides from erythrocytes (2) and thyroglobulin (23) have been shown to contain branched chain structures which have galactose as well as sialic acid or fucose as terminal sugar, while other glycoproteins such as ovalbumin (39) and rhodopsin (28) contain only N-acetylglucosamine and mannose. Thus it was important to establish if the two groups of glycopeptides, found in the synaptosomal plasma membrane, existed in the same protein or in different proteins. In order to consider this problem we have attempted to selectively solubilize the glycoproteins by extracting the membranes sequentially with various reagents and limiting quantities of different detergents.

As shown in table VIII aqueous solutions of high salt concentration, EDTA and mercaptoethanol, followed by urea solubilized about 30 percent of the proteins but failed to extract extensive amounts of protein-bound carbohydrate. However 1 percent Triton removed virtually all the sialic acid with proportionately lower amounts of mannose and galactose. Brunngraber et al. (10) have also observed that Triton selectively extracted glycoproteins from total particulate fractions of rat brain. A subsequent extraction with 0.1 percent sodium dodecyl sulphate removed about 15 percent of the proteins but proportionately higher amounts of mannose and galactose than sialic acid. Finally the pellet, which contained 15 percent of

TABLE VIII

Differential Extraction of Synaptosomal Membrane Carbohydrates

Sequential extraction Reagent	Component[1]			
	Protein	Sialic acid	Mannose	Galactose
1M NaCl;1mM EDTA;1mM MeSH	13.4	5.5	3.8	7.6
3M urea	16.7	2.8	2.2	6.5
1% Triton X-100	33.9	85.9	59.0	60.0
0.1% SDS	21.1	6.5	17.5	19.7
Insoluble residue	15.9	0.3	17.5	5.9

[1]Values are given as the percent of the total material in the sample.

the protein contained very little sialic acid and galactose but considerable amounts of mannose. This suggested that the sialic acid-rich glycoproteins were preferentially extracted before the mannose-rich glycoproteins.

In order to investigate this further we extracted a microsomal fraction (which also contains neuronal plasma membrane fragments) with Triton X-100 to obtain a Triton soluble and a Triton insoluble fraction. These two fractions were then analyzed for glycopeptides (Fig. 6.). It can be seen that virtually all of the sialic acid-

SEPHADEX G-50 CHROMATOGRAPHY OF THE GLYCOPEPTIDES FROM THE TRITON-SOLUBLE AND -INSOLUBLE MICROSOMAL FRACTIONS

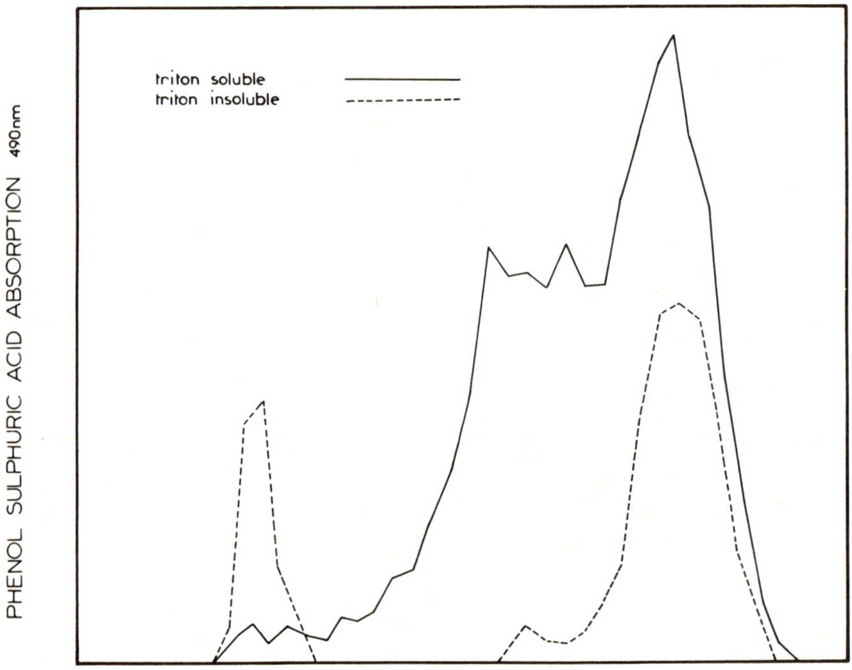

Fig. 6. Selective solubilization of microsomal membrane glycoproteins. ——— phenol sulphuric acid reaction after chromatography on Sephadex G-50 of glycopeptides produced by pronase hydrolysis of glycoproteins solubilized by Triton X-100. ------ phenol sulphuric acid reaction after chromatography on Sephadex G-50 of glycopeptides produced by pronase hydrolysis of glycoproteins remaining in the Triton insoluble pellet.

rich glycopeptides (S-I)were extracted with the Triton as well as a large quantity of the mannose-glucosamine rich species (S-II). However the Triton insoluble pellet contained only the S-II glycopeptides. These results suggest that the sialic acid-rich and mannose glucosamine-rich glycopeptides are derived in part from different peptide chains. Thus the latter species are not simply due to microheterogeneity within a given glycoprotein, provided that the microheterogeneity does not affect the solubility of the glycoprotein. The above results also suggest an obvious means of a partial fractionation of membrane proteins which could be readily combined with polyacrylamide gel electrophoresis.

Conclusion

The present studies have provided a partial characterization of the glycoproteins in synaptosomal plasma membranes. The membranes are rich in protein-bound carbohydrate (6-8 percent of protein). It has been reported (40) that there is a relatively limited number of glycoproteins in these membranes, while there are many other proteins. This suggests that the major glycoproteins will be considerably richer in carbohydrate than is the total membrane. It has been shown that two different types of carbohydrate chains are present in the glycoproteins. One group, which appears to be similar to glycopeptides of many well-defined soluble glycoproteins, contains sialic acid, fucose, galactose, mannose and glucosamine. As well there is a lower molecular weight group which contains mannose and glucosamine as the major components. The latter species appears to be partially associated with different glycoproteins than those glycopeptides containing sialic acid and fucose, since virtually all sialic acid glycopeptides are extracted by Triton leaving a residue which is rich in glycopeptides containing mannose and glucosamine. It is possible that there are glycoproteins present in the neuronal plasma membrane which contain carbohydrate chains of glucosamine and mannose with a structure similar to the carbohydrate chains found in such soluble glycoproteins as ovalbumin (39) and rhodopsin (28).

Although no specific function is obvious at the present time it is possible to envisage potential roles for these glycoproteins. It has been claimed that after Triton treatment of synaptosomes the synaptic complex is left partially intact (22). It is possible that specific glycoproteins are present in this region and could be responsible for close intercellular contacts which may exist in the synaptic complex. Such carbohydrate chains could also serve as suitable precursors for the addition of other carbohyrates as has been claimed by Barondes (3, 4). This suggestion, however, needs to be carefully checked by well-controlled tests for the presence of glycosyl transferases in these highly purified membranes. Initial studies (M. Reith, unpublished observations) indicate that galactosyl transferase is not enriched in these membranes in comparison to the total homogenate.

In conclusion, the synaptosomal plasma membrane contains glycoproteins with carbohydrate chains which appear to be able to present many flexible characteristics that would be useful for intracellular contacts or modification of the neuronal surface.Further studies on their isolation, characterization and localization in the synaptosomal plasma membrane should provide additional information concerning their possible functions.

ACKNOWLEDGEMENTS

W.C.Breckenridge : Fellow of the Medical Research Council of Canada. I.G.Morgan : Attaché de Recherche au CNRS.

This work has been supported by a grant from the Institut National de la Santé et de la Recherche Médicale (contract no. 71 116 98). We thank Drs. G.Gombos, L.S.Wolfe and G.Vincendon for their assistance and support during the performance of this work.

REFERENCES

1 ABERCOMBIE, M. and HEARPAMAN, J., Exptl. Cell Res., 6 (1954)293.
2 ADAMANY, A.M. and KATHAN, R.H., Biochem. Biophys. Res. Commun., 37 (1969) 171.
3 BARONDES, S.H., in The Neurosciences : Second Study Program (Ed. SCHMITT, F.A.) The Rockefeller University Press, New York,1970, p.747.
4 BARONDES, S.H., J. Neurochem., 15 (1968) 699.
5 BEELEY, J.G., Biochem. J., 123 (1971) 399.
6 BOSMANN, H.B., HAGOPIAN, A. and EYLAR, E.H., Arch. Biochem. Biophys., 128 (1968) 51.
7 BOGOCH, S., Neurosci. Res. Progr. Bull., 3 (1965) 38.
8 BRECKENRIDGE, W.C., BRECKENRIDGE, J.E., VINCENDON, G. and GOMBOS, G., in preparation.
9 BRECKENRIDGE, W.C., GOMBOS, G., VINCENDON, G. and MORGAN, I.G., in preparation.
10 BRUNNGRABER, E.G., AGUILAR, V. and ARO, A., Arch. Biochem. Biophys., 129 (1969) 131.
11 BRUNNGRABER, E.G., ARO, A. and BROWN, B.D., Clin. Chim. Acta, 29 (1970) 333.
12 BRUNNGRABER, E.G., BROWN, B.D. and AGUILAR, V., J. Neurochem., 16 (1969) 1059.
13 BRUNNGRABER, E.G., in Handbook of Neurochemistry (Ed. LAJTHA,A.) Plenum Press, New York, 1969, vol. 1, p.223.

14 CARLSON, D.M., J. Biol. Chem., 243 (1968) 616.
15 QUARLES, R.H. and BRADY, R.O., J. Neurochem., 17 (1970) 801.
16 DE BERNARD, B., PUBLIARELLO, M.C., SANDRI, G., SOTTOCASA, G.L. and VITTUR, F., Febs Letters, 12 (1971) 125.
17 DEKIRMENJIAN, H. and BRUNNGRABER, E.G., Biochim. Biophys.Acta, 177 (1969) 1.
18 DI BENEDETTA, G., BRUNNGRABER, E.G., WHITNEY, G., BROWN, B.D. and ARO, A., Arch. Biochim. Biophys., 131 (1969) 404.
19 DUNN, J.T. and SPIRO, R.G., J. Biol. Chem., 242 (1967) 5556.
20 EMMELOT, P., BOS, C.J., BENEDETTI, E.L. and RUMKE, Ph., Biochim. Biophys. Acta, 90 (1964) 126.
21 EVANS, W.H., Biochim. Biophys. Acta, 211 (1964) 578.
22 FISZER, S. and DE ROBERTIS, E., Brain Res., 5 (1967) 31.
23 FUKUDA, M. and EGAMI, F., Biochem. J., 123 (1971) 407.
24 GLICK, M.C., COMSTOCK, C. and WARREN, L., Biochim. Biophys. Acta 219 (1970) 290.
25 GINSBURG, V. and NEUFELD, E.F., Ann. Rev. Biochem., 38 (1969) 371.
26 GOTTSCHALK, A., Nature, 222 (1969) 452.
27 HEILJMAN, J. and ROUKEMA, P.A., Biochim. Biophys. Acta, 127 (1966) 269.
28 HELLER, J. and LAWRENCE, M.A., Biochemistry, 9 (1971) 864.
29 HENNING, R., KARULEN, H.D. and STOFFEL, W., Hoppe Seyler Z. Physiol. Chem., 351 (1970) 1191.
30 LLOYD, K.O., KABAT, E.A. and LICERIO, E., Biochemistry, 7 (1968) 2976.
31 MIYAJIMA, N., KAWASAKI, T. and YAMASHINA, I., Febs Letters, 11 (1970) 29.
32 MOCZAR, E., MOCZAR, M., SCHILLINGER, G. and ROBERT, L., J. Chromatog., 31 (1967) 561.
33 MOLNAR, J., MARKOVIC, G. and CHAO, H., Arch. Biochem. Biophys., 134 (1969) 524.
34 MORGAN, I.G., WOLFE, L.S., MANDEL, P. and GOMBOS, G., Biochem. Biophys. Acta, 241 (1971) 737.
35 RAMBOURG, A. and LEBLOND, C.P., J. Cell Biol., 32 (1967) 27.
36 SCOTT, J.E., in Methods Biochem. Anal., 8 (1960) 145.
37 SHIMIZU, S. and FUNAKSOHI, I., Biochim. Biophys. Acta, 203 (1970) 167.
38 SHIMIZU, A., PUTNAM, F.W., RAUL, C., CLAMP, J.R. and JOHNSON, I., Nature New Biol., 231 (1971) 73.
39 SPIRO, R.G., Ann. Rev. Biochem., 39 (1970) 599.
40 WAEHNELDT, T.V., MORGAN, I.G. and GOMBOS, G., Brain Res., 34 (1971) 403.
41 WARREN, L., J. Biol. Chem., 234 (1959) 1971.
42 WHITTAKER, V.P., MICHAELSON, I.A. and KIRKLAND, R.J.A., Biochem. J., 90 (1964) 293.
43 WOLFE, L.S., GOMBOS, G., BRECKENRIDGE, W.C., MANDEL, P. and MORGAN, I.G., in preparation.

SECTION 4

MYELIN PROTEINS

SEPARATION AND CHARACTERIZATION OF MYELIN PROTEINS

E. MEHL

Max-Planck-Institut of Psychiatry, Division of Neurochemistry - Munich (Germany)

Within the group of plasma membranes, isolated and studied until now, the myelin membrane has a typical characteristically low protein content of 20 percent (3,15,18) and a protein composition of reduced complexity. By work on isolated cerebral myelin, the encephalitogenic basic protein was directly shown to be a myelin protein (17, 18). Autilio (2) applied gel filtration in chloroform-methanol and obtained a further myelin protein fraction comprising 55 to 65 percent of total myelin proteins. This protein fraction had an amino acid composition almost identical to the classical Folch-Lees proteolipid protein (13) which was extracted from cerebral white matter and subjected to a purification step.

As a more rigid criterion of homogeneity than gel filtration, polyacrylamide gel electrophoresis with phenol-formic acid-water, instead of aqueous buffers as conducting medium, was applied.Under this condition, myelin was nearly completely dissolved and separated into about ten distinct protein zones. The Folch-Lees proteolipid protein appeared as a narrow zone comprising 50 percent of the total proteins and it was also present in cerebral myelin of rat and carp despite their atypical basic proteins (23, 24). The low molecular weight at 35,000 as determined by gel electrophoresis across linear gradients of polyacrylamide with the phenolic solvent as conducting medium could indicate that the proteolipid protein migrated as a monomer instead of a composite complex of different subunits (30). The presence of a third cerebral myelin protein fraction especially rich in glutamic acid was indicated, as the content of this amino acid in the total protein hydrolyzate of myelin was too high to be accounted for by any possible combination of the basic protein fraction and the Folch-Lees proteolipid protein (35). The Triton-salt fractionation technique (10) is suitable for removal of this Wolfgram protein fraction as Triton-salt insoluble residue and leads to a precipitation of the extracted proteolipid fraction by depleting the Triton with ether, leaving the basic protein still soluble.

By means of polyacrylamide gel electrophoresis in the phenolic solvent, three myelin types could be distinguished in the bovine nervous system on the basis of their typical protein components (25). The stoichiometric relations between the main proteins were determined (24). Some extensions of our previous work (23-25,30) and the advantages of recent methods used by other groups are the contents of the report.

The purity of isolated myelin fractions has been studied by electron microscopy (18, 24), RNA analysis (3, 24) and determination of enzyme activities (4, 18). Phenol- and SDS-polyacrylamide gel electrophoresis (31) provide further criteria of its purity. After gel electrophoresis in phenol-formic acid-water (14:3:3, w/v/v), the extent of purification was apparent. The densitometric tracing (Fig. 1) of myelin protein pattern indicated 10 zones in contrast to the 25 protein zones of the sample of cerebral white matter. The

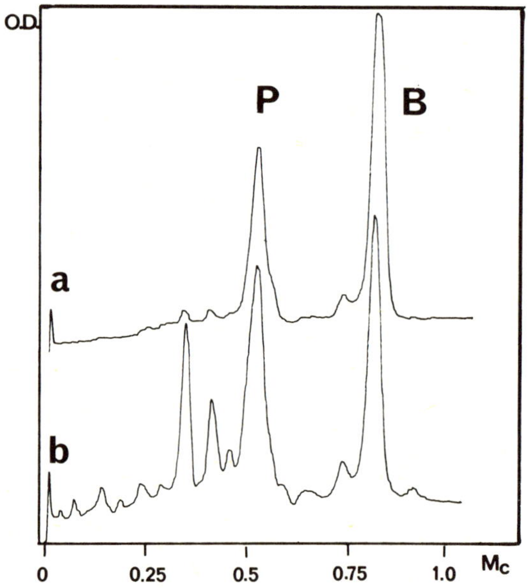

Fig. 1. Purity of isolated myelin. - Densitometric tracing of protein patterns of myelin (a) isolated from bovine cerebral white matter (b). For gel electrophoresis a 7.5 % polyacrylamide gel with phenol-formic acid-water (14:3:3, w/v/v) was used. Mc, mobility relative to cytochrome c ; B, myelin basic protein ; P, Folch-Lees proteolipid protein.

two main proteins were identified as myelin basic protein and Folch-Lees proteolipid protein upon co-electrophoresis with purified reference samples (basic protein 16, proteolipid protein 6). The multiple minor protein zones of the myelin fraction are the Wolfgram-type proteins, as will be seen later. To study the question, whether these minor proteins were myelin proteins or constituents of a contaminating particulate fraction, myelin was prepared by zonal rotor centrifugation in a continuous sucrose gradient (8). The protein composition of the subfractions was studied by means of SDS-gel electrophoresis (31) under modified conditions of sample dissolution (20 mg of SDS per mg of protein). The gels were stained according to Weber and Osborn (32). Regarding the possible source of Wolfgram-type protein, indication was obtained that myelin is contaminated by a particulate fraction which critically resembles myelin in its sedimentation characteristics. The concentration of the pair of minor zones (Fig. 2) varied independently of the occurence of the

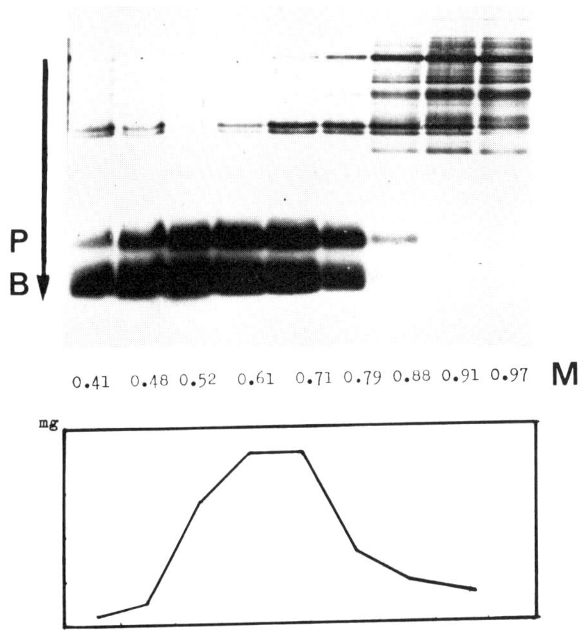

Fig. 2. Variability of minor protein zones in myelin subfractions.- Myelin subfractions were prepared from 40 g of rabbit brain in a continuous sucrose gradient by zonal rotor centrifugation after osmotic shock (8). The sucrose concentration of the various subfractions is expressed as Mole/litre. Lower half, profile of protein concentration of the gradient fractions. The subfractions were compared by SDS (sodium dodecylsulphate) gel electrophoresis (31) on a gel slab (15 g acrylamide and 0.2 g BIS per 100 ml). As a modification, the sample solution contained 2 % SDS and 0.2 % of protein. 10 μl samples were applied. B, basic protein ; P, proteolipid protein.

highly concentrated proteolipid protein plus basic protein. The Wolfgram-type proteins were barely detectable at 0.52 M-sucrose. Their concentration increased in such a way that they became the major constituents at 0.88 M-sucrose, whereas the typical myelin proteins were fading out.

A further question was, whether some of the Wolfgram-type proteins might have escaped gel electrophoretic analysis because of incomplete solubilization and incomplete disaggregation of the myelin proteins. For a radiochemical approach, myelin was labelled with L-leucine-1-^{14}C (62 mCi/mmol). A dose of 30 μCi was injected intracerebrally to 3 week-old rabbits. The phenol-formic acid-water (14:3:3, w/v/v) solvent solubilized labelled myelin up to 97.5 per cent at a protein concentration of 1 mg per ml, as judged by the ^{14}C-activity remaining in the supernatant upon 3 h of centrifugation at 150,000 g at 20°. Up to 82 percent of the ^{14}C-activity were found in the supernatant, when myelin was solubilized with 20 mg SDS per mg of protein contained in one ml. Only 40 percent of the activity was left in the supernatant, when we followed the original method (31) using 0.6 mg of protein plus 1 mg SDS per ml. In addition to the 2.5 percent of ^{14}C-labelled proteins, insoluble in the phenolic solvent, a further 2 percent (11 counts/min of 600 counts/min applied) were found left at the filter paper(sample origin) upon completion of electrophoresis, and 3 percent in the unstained 2 mm-layer of large pore gel.

Cerebral myelin samples of various species were subjected to gel electrophoresis in the phenol solvent (Fig. 3).The relative

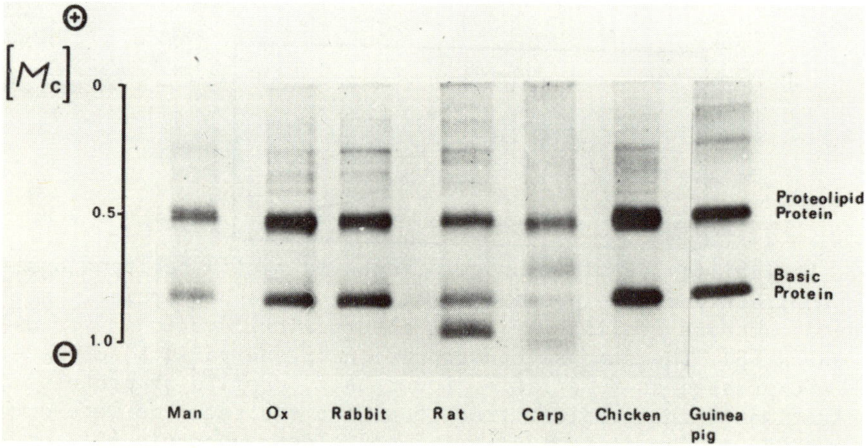

Fig. 3. Electrophoretic protein pattern of cerebral myelin from various species. Myelin samples containing about 0.02 mg protein were separated on a 7 % polyacrylamide gel with phenol-formic acid-water (14:3:3, w/v/v).

mobility of each zone was expressed as a M_c value. This value represents the ratio of the distance of zone from sample origin to the distance of the cytochrome c marker from sample origin (5).The myelin samples of all the species contained the Folch-Lees proteolipid (M_c 0.54) as a main constituent and also the basic protein (M_c 0.82), except for the carp where the basic protein was replaced by two protein zones (M_c 0.71 and M_c 1.00). In the myelin from rat brain, a second basic protein (M_c 0.95) was present, which was independently established with acid extracts of rat brain (22). These two basic proteins of the rat were purified and the amino acid composition was determined (19). Since these atypical basic proteins were also detectable in samples of freshly dissolved cerebral white matter, their appearance as a result of proteolytic activity during the myelin isolation procedure is unlikely. Densitometric evaluations of the electrophoretic patterns similar to those in Fig. 1. were carried out. After gel electrophoresis of isolated reference samples of basic protein and proteolipid protein, both proteins bound the same amount of stain per mg of nitrogen within $\pm$ 12 percent. According to densitometric evaluations of cerebral myelin samples from five vertebrate species including chicken, a constant relation of 1.0 : 0.690 ($\pm$ 0.013) : 0.4 was found for proteolipid protein, basic protein, heterogeneous fraction of minor protein zones, respectively. Upon the Triton-salt fractionation technique and amino acid analysis, the relation was 1.0 : 0.6 : 0.4 for proteolipid protein, basic protein, Wolfgram-type proteins, respectively, in human and bovine cerebral myelin (10). This agreement of results is one argument that the minor protein zones are Wolfgram-type proteins. When examined by SDS gel electrophoresis, the Wolfgram-type protein fraction (Triton-salt insoluble residue) was indeed heterogeneous and consisted of components of higher molecular weight.

In contrast to the general constancy of the protein composition of cerebral myelin, whether mammalian or chicken,different myelin types exist at different anatomical sites in the bovine nervous system (Fig. 4, 5, 6). An additional protein (M_c 0.47) was present in myelin from the spinal cord with a corresponding reduction in the amount of proteolipid protein (M_c 0.52). The concentration of the proteolipid protein was found to be reduced in the anterolateral columns of spinal cord (1) in comparison to cerebral white matter, and differences in amino acid composition were established between cerebral and spinal myelin (35). Peripheral nerve myelin was prepared from the spinal roots (34). The main proteins (M_c 0.72 and M_c 0.91) were completely different from those of cerebral myelin. The Folch-Lees proteolipid protein was not detectable. The Triton-salt insoluble residue contained 50 percent of the proteins (10). Wolfgram and Kotorii (34) established by acid extraction, proteolytic treatment and amino acid analysis, that the myelin basic protein was absent and that the Folch-Lees proteolipid protein was

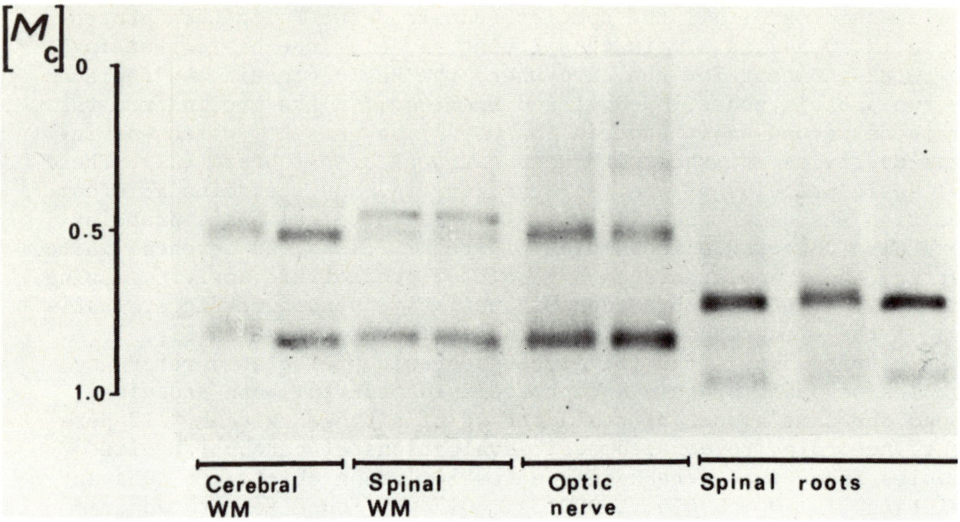

Fig. 4. Electrophoretic protein pattern of bovine myelin types. Myelin was isolated from cerebral white matter, spinal cord white matter,optic nerve and spinal roots (representing peripheral nerve). For conditions see Fig. 1.

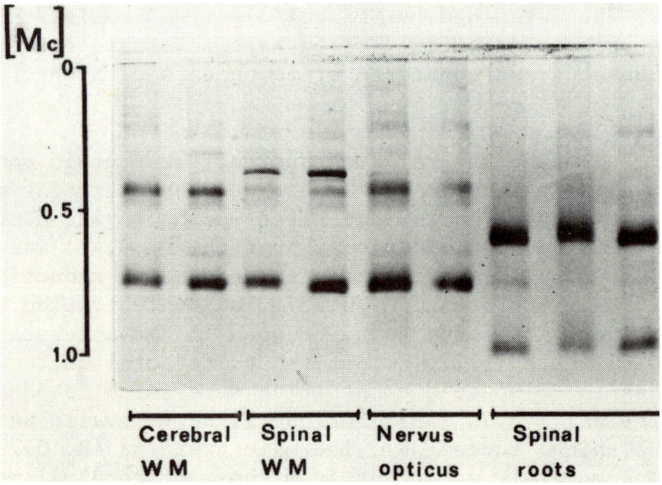

Fig. 5. Electrophoretic protein pattern of isolated bovine myelin types. The 12 % gel slab was equilibrated with phenol-acetic acid-6 M urea (2:1:1, w/v/v) pH 2.0 before electrophoresis. A similar solvent has been used for electrophoretic separation of myelin proteins of the rat (7).

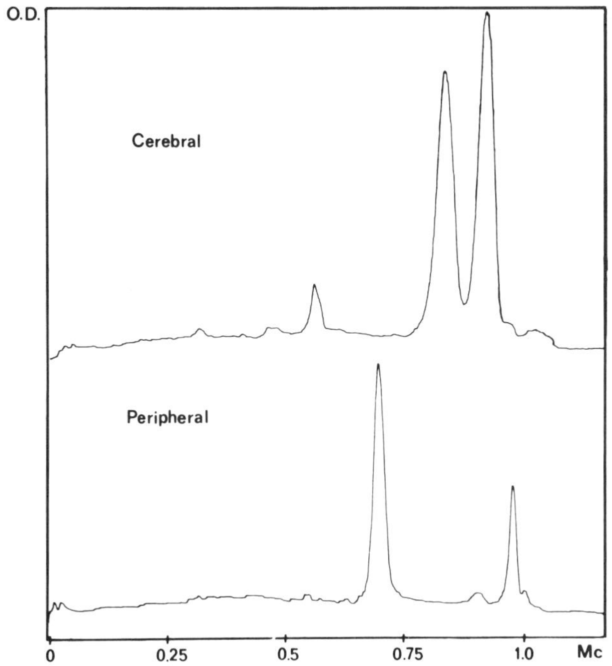

Fig. 6. Densitometric tracing of bovine cerebral myelin proteins and tracing of myelin proteins from bovine peripheral nerve (spinal roots). SDS gel electrophoresis was performed as described (Fig.2).

replaced in part by another type of protein. The results (Fig. 4) were confirmed when gel and samples contained phenol-acetic acid-6 M urea (2:1:1, w/v/v) (Ref. 7) (Fig. 5) or SDS (Fig. 6).

The relative amounts of the major protein components were determined by densitometric quantitation of the stained protein zones upon gel electrophoresis in the phenol solvent. The molecular weights were determined by gel electrophoresis in phenol-formic acid-water (Ref. 24) and phenol-acetic acid-water (Ref. 30) or in SDS gels (Table I). The molecular weight of the basic protein (SDS) is in accordance with the established data (27) in contrast to the figure in the phenolic system. The molecular weights of the main components of peripheral nerve myelin agree reasonably closely in the two systems. Regarding Folch-Lees proteolipid protein, the molecular weights of 35,000 (phenol) and 23,000 (SDS) could be explained (14) by a trimer- and a dimer-state with the molecular weight of the monomer at 12,000. On the basis of amino acid analysis, this figure has been predicted as the minimum molecular weight (12). The figures (Table I) are closely in line with a 1:1 stoichiometry of basic protein and proteolipid protein. A high degree of order in the supramolecular structure of myelin membrane is further evidenced

TABLE I

Stoichiometry of the Two Main Protein Components in Myelin from Cerebral White Matter and from Bovine Peripheral Nerve

Myelin proteins		Molecular weight SDS-method†	Molecular weight Phenol method	Relative amount
Cerebral white :				
Proteolipid protein	A	23,000	35,000††	1
Basic protein	B	18,300	25,000††	0.705††
Molar ratio (A : B)		1 : 0.89	1 : 1	
Peripheral nerve :				
Protein	C	28,200	28,000	1
Protein	D	14,700	17,000	0.45
Molar ratio (C : D)		1 : 0.86	1 : 0.74	

†Method of Weber and Osborn (32). ††Average for ox and guinea pig.

by the constant quantitative relation between these two proteins in five species. If the monomer of 12,000 would exist, a 2:1 stoichiometry follows.

The 1 : 1 stoichiometric relation was taken as an indication that the intact myelin membrane might contain heterodimers of the basic protein and the proteolipid protein as the two subunits. To capture this dimer of molecular weight 41,000 (18,000 + 23,000), cerebral myelin, 1 mg protein/ml, was reacted with the cross-linking reagent N-ethyl-N'(3-dimethylaminopropyl) carbodiimide HCl (K&K Lab. Plainview, N.Y.) for 60 h at 4°. Six mg of reagent were used per ml of 20 mM phosphate buffer (pH 7.4). After dissolution of cross-linked samples along with control samples in SDS, such a dimer was not detectable by means of SDS gel electrophoresis (Fig. 7). In sample B, the basic protein (zone 2) was not detectable and nearly completely prevented from entering the gel, as indicated by an arrow at the origin. This polymerization of basic protein was not detectable, when myelin was dissolved in SDS before addition of the reagent (control A). Proteolytic loss of basic protein was excluded by the control C, which was maintained at 4° as long as B. Reaction products of the reagent were stained with Coomassie blue (zone 3). This polymerization of basic protein indicated that the basic protein molecules contact each other in the intact membrane.

What are the arguments for homogeneity of the zones as separated by means of gel electrophoresis with phenol-formic acid-

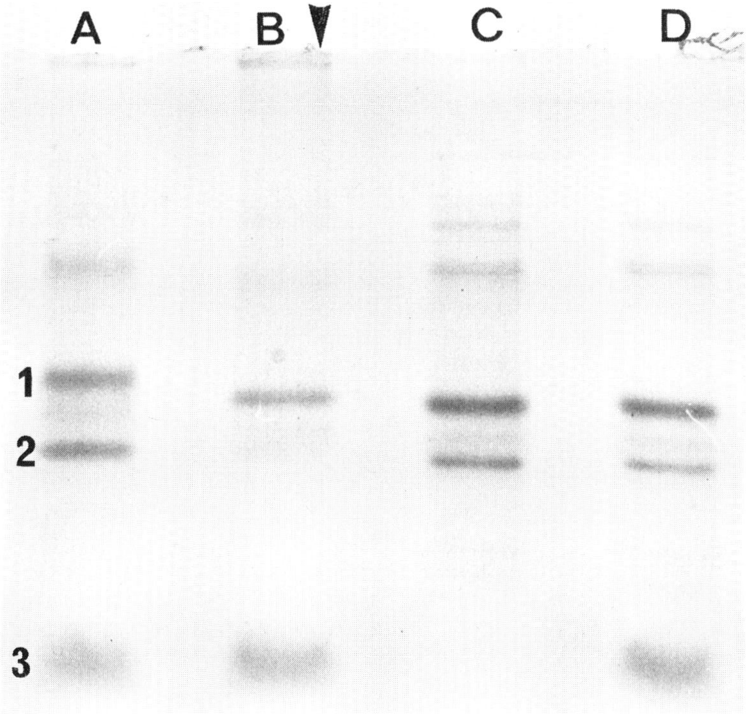

Fig. 7. Effect of cross-linking reagent on the intact myelin membrane. Myelin samples were reacted with a cross-linking reagent for 60 h at 4° and subjected to SDS gel electrophoresis. Control A, myelin was dissolved with 2 % SDS prior to the addition of reagent; B, myelin sample with cross-linked basic protein ; control C, myelin was incubated without reagent as a control for autolysis ; control D, mixture of C and 60 h-old reagent solution as a control of unaffected solubility in SDS. For further conditions see text.

water? The use of gel electrophoresis with this solvent for determination of molecular weights established that separation is mainly based on differences in molecular size (30). The acidic conditions which were necessary for solubilization of myelin proteins might obscure minor charge differences which would be detectable under electrophoretic conditions nearer to the isoelectric point of the protein. Myelin basic protein separated as one zone at pH 2.4 and was resolved into a fast and a slow zone when electrophoresed at pH 10.6 (21). The fast component was non-enzymatically transformed into the more slowly migrating zone depending on the conditions of storage in chloroform-methanol (20). To obtain additional arguments for the homogeneity of the zones as they appear after gel electrophoresis in the phenol-formic acid system, the proteins were isolated in that solvent system.

Molecular exclusion chromatography, using Biogel P-10 (200 - 400 mesh, exclusion limit 10,000, Bio-Rad Laboratories) and the solvent phenol-98 percent (w/w) formic acid-water (14:3:3, w/v/v), were applied for the preparative separation of myelin into proteins and lipids. After washing a batch of Biogel with 0.2 M-ammonium sulphate solution containing 0.2 M - NaCl, the gel was washed with distilled water until free of chloride ions. The Biogel was dehydrated with methanol and dried under reduced pressure. After soaking in the phenolic solvent overnight, the gel was packed into a chromatography column (2.5 x 42 cm). A 100-200 mg sample of the freeze-dried pig myelin was dissolved in 10 ml of the phenolic solvent.The solution passed through the column at a rate of 10 ml per hour and the main protein peak appeared when 60 ml of effluent had left the column. Fifty ml of protein-containing effluent were mixed with 54 ml of 50 percent (v/v) acetic acid and 52 ml of benzene. The purified protein was in the benzene-rich upper phase and the basic protein in the lower one. No insoluble material was detectable at the interphase. For further purification of the proteolipid protein, the upper phase was washed with a freshly prepared (protein-free) lower phase. As studied with bovine myelin, the myelin proteins were separated from the lipids after the step of gel filtration in the phenolic solvent (Fig. 8). The protein fraction contained less than 0.1 % phosphorus per mg of protein. This result made it

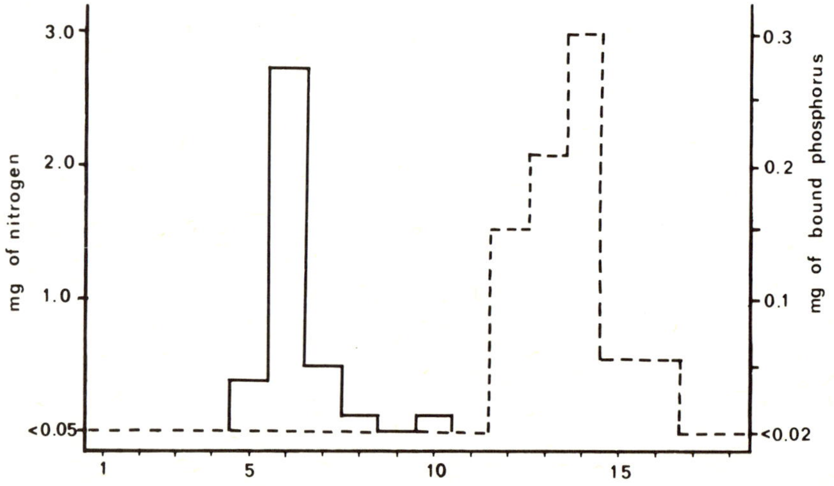

Fig. 8. Molecular exclusion chromatography of freeze-dried bovine myelin. The solvent was phenol-98 % (w/w) formic acid-water (14:3:3, w/v/v). 10 ml of a 1 percent (w/v) solution of myelin were passed through a column (2.5 x 42 cm) of Biogel P-10 (nominal exclusion molecular weight 10,000). The fraction volume was 12 ml. Myelin proteins in nos. 5-8 ; phospholipids in nos. 12-16. The phenolic solvent dissociates the myelin into its proteins and lipids.

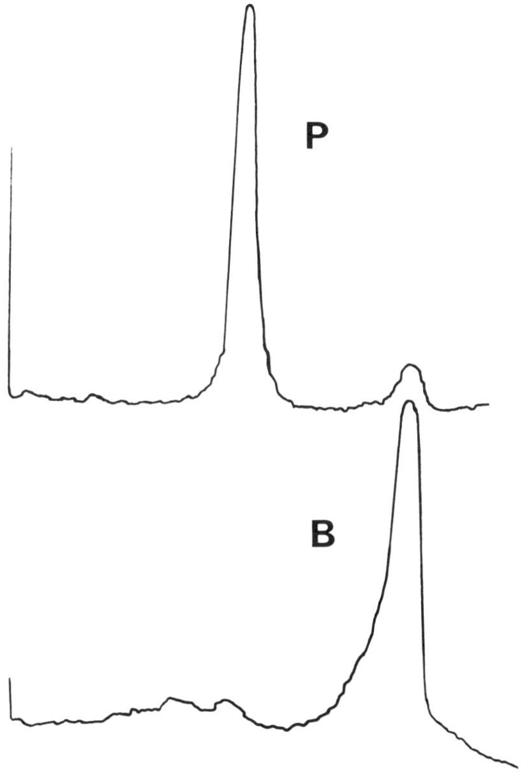

Fig. 9. Test of homogeneity of myelin proteins, after isolation by phenol-formic acid-benzene distribution. Densitometric tracing of Folch-Lees proteolipid protein (upper phase,P)and the basic protein plus Wolfgram-type proteins (lower phase, B). Samples of 20 µl were separated by gel electrophoresis with the phenolic solvent.

possible to electrophorese dissolved myelin samples without any preceding lipid extraction. The homogeneity of the isolated Folch-Lees proteolipid protein (benzene-rich upper phase) and of the basic protein (lower phase) was tested by analytical electrophoresis on aliquots of the phases before lyophilization (Fig. 9). The proteolipid protein completely penetrated into the gel as one zone corresponding to a molecular weight of 35,000. The protein was nearly quantitatively distributed into the benzene-rich upper phase, thus behaving as homogeneous proteolipid protein. The amino acid composition was essentially the same as that of other proteolipid protein preparations (Table II). Therefore, the main zone (M_c 0.54) was indeed the Folch-Lees proteolipid protein. The lower phase

TABLE II

Comparison of the Amino Acid Composition (Mole %) of the Folch-Lees Proteolipid Protein after Various Preparation Techniques

Source	Cerebral white matter (Bovine) A	Cerebral white matter (Bovine) B	Myelin (Pig) C	Myelin (Bovine) D
Lysine	4.5	4.3	4.4	4.8
Histidine	2.3	1.9	1.9	2.3
Arginine	2.6	2.6	2.2	3.0
Aspartic	4.0	4.2	3.7	4.4
Threonine	8.4	8.5	8.9	8.6
Serine	5.2	5.4	6.5	6.7
Glutamic	5.9	6.0	6.1	6.6
Proline	2.8	2.9	2.5	2.3
Glycine	10.4	10.3	10.1	11.0
Alanine	11.9	12.5	11.1	11.8
Valine	7.4	6.9	5.9	7.0
Half-cystine+	2.9	4.2	5.7	2.4
Methionine	1.3	1.7	1.5	0.5
Isoleucine	5.1	4.9	5.0	5.3
Leucine	11.5	11.1	11.8	11.1
Tyrosine	4.4	4.7	4.7	4.0
Phenylalanine	8.2	7.8	8.1	8.0

+Uncorrected for degradation caused by acid hydrolysis. A) data from Ref. 35 ; B) data from Ref. 29 ; C) prepared by phenol-acid - benzene distribution (hydrolysis for 20 h at 110° with 6 N-HCl in vacuum-sealed tubes, Beckman-Spinco model 120 amino acid analyzer); D) prepared by the Triton-salt technique. Data from Ref. 10.

contained myelin basic protein plus Wolfgram-type protein zones. In comparison to isolated myelin basic protein (26) this purified fraction contained only 20 percent instead of 25 percent of basic amino acids.

The reliability of the analytical polyacrylamide gel electrophoresis of myelin proteins with phenol-formic acid-water (14:3:3, w/v/v) as buffer was thus tested under various aspects.The method was used to study proteolytic changes of myelin proteins during demyelination (28). A modified technique was applied for analysis of the occurence of myelin proteins during the period of myelination (9). The proteolipid protein obtained by the Triton-salt technique behaved as a homogeneous protein in phenol-formic acid-water (2:1:1, w/v/v) gels (11). This modified electrophoretic system had a disadvantage, since the proteolipid protein migrates as an aggregated compound of a molecular weight comparable to serum albumin. Replacement of phenol-formic acid by trichloroethanol-formic acid for gel filtration and polyacrylamide gel electrophoresis has the advantage that analytical as well as preparative polyacrylamide gels can be prepared in the presence of the organic solvent (33).

Summarizing for isolated cerebral myelin, the Folch-Lees proteolipid protein accounts for 50 percent of the proteins as a homogeneous component, in both SDS- and phenol-formic acid-polyacrylamide gel electrophoresis systems. The molecular weight in the former and the latter system was determined as 23,000 and 35,000 respectively. Excepting the unusual basic proteins of rat and carp, 1 mole of basic proteins occurs together with 1 mole of proteolipid protein in cerebral myelin membrane. This indication of basic protein and proteolipid protein as subunits of a dimer complex could not yet be demonstrated by cross-linking reagent techniques. Basic protein and proteolipid protein partition as homogeneous proteins, when purified in a two-phase system of phenol-formic acid-benzene. The occurence of additional minor proteins in myelin is variable and depends greatly on the purity of the isolated myelin.

In spinal myelin, an additional proteolipid protein of higher molecular weight is present with a corresponding reduction in the amount of Folch-Lees proteolipid protein.

According to the molecular weight determination in the two gel-electrophoretic systems, peripheral nerve myelin contains a 28,000-protein as main component and a 15,000-protein.

Cross-linkage experiments indicate that the basic protein molecules contact each other in the membrane of cerebral myelin.

Reference is made to semi-preparative gel electrophoretic methods and the advantages of the Triton-salt fractionation technique.

REFERENCES

1 AMADUCCI, L., PAZZAGLI, A. and PESSINA, G., J. Neurochem., 9 (1962) 509.
2 AUTILIO, L., Feder. Proc., 25 (1966) 764.
3 AUTILIO, L.A., NORTON, W.T. and TERRY, R.D., J. Neurochem., 11 (1964) 17.
4 BANIK, N.L. and DAVISON, A.N., Biochem. J., 115 (1969) 1051.
5 CARNEGIE, P.R., LAMOUREUX, G. and BENCINA, B., Nature (Lond.), 214 (1967) 407.

6 CAVANNA, R. and RAPPORT, M.M., J. Lipid Res., 8 (1967) 65.
7 COTMAN, C.W. and MAHLER, H.R., Arch. Biochem., 120 (1967) 384.
8 COTMAN, C.W., MAHLER, H.R., and ANDERSON, N.G., Biochim.Biophys. Acta, 163 (1968) 275.
9 EINSTEIN, E.R., DALAL, K.B. and CSEJTEY, J., Brain Res., 18 (1970) 35.
10 ENG, L.F., CHAO, F.C., GERSTL, B., PRATT, S. and TAVASTSTJERNA, M.G., Biochemistry, 7 (1968) 4455.
11 ENG, L.F. and coworkers, J. Neurobiol., in press.
12 FOLCH-PI, J., in Brain Lipids and Lipoproteins and the Leucodystrophies (Ed. FOLCH-PI, J. and BAUER, H.J.) Elsevier Publ. Co., Amsterdam, 1963, p.18.
13 FOLCH, J. and LEES, M., J. Biol. Chem., 191 (1951) 807.
14 FOLCH-PI, J., 3rd Meeting Intern. Soc. Neurochem., Budapest, 1971, Abstr. p.413.
15 HULCHER, F.H., Arch. Biochem., 100 (1963) 237.
16 KIES, M.W., Ann. N.Y. Acad. Sci., 122 (1965) 161.
17 KIES, M.W., THOMPSON, E.B. and ALVORD, E.C., Ann. N.Y. Acad. Sci., 122 (1965) 148.
18 LAATSCH, R.H., KIES, M.W., GORDON, S. and ALVORD, E.C., J. Exptl. Med., 115 (1962) 777.
19 MARTENSON, R.E., DEIBLER, G.E. and KIES, M.W., Biochim. Biophys. Acta, 200 (1970) 353.
20 MARTENSON, R.E., DEIBLER, G.E. and KIES, M.W., J. Neurochem., 17 (1970) 1329.
21 MARTENSON, R.E. and GAITONDE, M.K., J. Neurochem., 16 (1969) 889.
22 MARTENSON, R.E., GAITONDE, M.K. and RICHTER, D., 1st Meeting Intern. Soc. Neurochem., Strasbourg, 1967, Abstr. p.151.
23 MEHL, E., 1st Meeting Intern. Soc. Neurochem., Strasbourg,1967, Abstr. p.154.
24 MEHL, E. and HALARIS, A., J. Neurochem., 17 (1970) 659.
25 MEHL, E. and WOLFGRAM, F., J. Neurochem., 16 (1969) 1091.
26 NAKAO, A., DAVIS, W.J. and EINSTEIN, E.R., Biochim. Biophys. Acta, 130 (1966) 163.
27 OSHIRO, Y. and EYLAR, E.H., Arch. Biochem. Biophys., 138 (1970) 606.
28 RIEKKINEN, P.J., PALO, J., ARSTILO, A.U., SAVOLAINEN, H.J., RINNE, U.K., KIVALO, E.K. and FREY, H., Arch. Neurol., 24 (1971) 545.
29 TENENBAUM, D. and FOLCH-PI, J., Biochim. Biophys. Acta, 115 (1966) 141.
30 THORUN, W. and MEHL, E., Biochim. Biophys. Acta, 160 (1968) 132.
31 WAEHNELDT, T.V. and MANDEL, P., FEBS Letters, 9 (1970) 209.
32 WEBER, K. and OSBORN, M., J. Biol. Chem., 244 (1969) 4406.
33 WOLFGRAM, F., 2nd Meeting Intern. Soc. Neurochem., Milano, 1969, Abstr. p.419.
34 WOLFGRAM, F. and KOTORII, K., J. Neurochem., 15 (1968) 1291.
35 WOLFGRAM, F. and KOTORII, K., J. Neurochem., 15 (1968) 1281.

PROTEOLIPIDS

J. FOLCH-PI

Harvard Medical School, Boston, Mass. 02115 and
McLean Hospital, Belmont, Mass. 02178 (USA)

By comparison to the large amount of work that has been carried out on myelin lipids in health and in disease, scanty attention has been paid until recently to myelin proteins. Indeed what was known about central myelin proteins in the late fifties had been surmised from the differences in composition between central gray and white matters, and from the changes in composition during maturation of the central nervous system. It could have been summarized by stating that neurokeratin (8), and the then recently recognized proteolipids (12) were most likely components of central myelin. Neurokeratin was the gastric juice, pancreatic juice resistant fraction of defatted brain proteins. Its occurrence had been known for almost a century, and in the fifties it had been prepared by improved methods as a trypsin-resistant protein residue (TRPR) (24). Proteolipids were a group of protein-lipid complexes that had been isolated from brain tissue in 1951, and which had the property of being insoluble in aqueous solutions, and soluble in chloroform-methanol mixtures (12). TRPR and proteolipids were obtained in parallel from a single tissue sample by extraction with chloroform-methanol : the extract contained the proteolipids, and TRPR was obtained by tryptic digestion of the chloroform-methanol insoluble tissue residue. The two materials were similar in amino acid composition, in their insolubility in aqueous solutions, and in being resistant to the action of most proteolytic enzymes. This left, as main difference between the two, the insolubility of TRPR in chloroform-methanol as compared to the ready solubility of proteolipids in the same solvent mixtures.

Peripheral nervous system myelin was known to contain neurokeratin, but only very small amounts of proteolipids (9, 13).

The isolation of central myelin by subcellular fractionation spurred the study of myelin proteins. The unexpected finding that isolated central myelin was almost completely soluble in chloroform-methanol (3,22) suggested originally that all of its protein was proteolipid in nature. A number of observations, however, soon indicated that the solubility of myelin in chloroform-methanol was misleading and that proteins other than proteolipids were present in myelin. Thus, Laatsch et al. (22) isolated from it a basic antigenic protein which produced allergic encephalomyelitis. This protein was quite different from proteolipids. Autilio (2), by permeation on polystyrene gels, was able to separate from central myelin, a proteolipid fraction, and a second fraction with the amino acid composition of the basic antigenic protein. In addition, Lees (26) demonstrated that when brain tissue homogenates were freed of diffusable electrolytes, chloroform-methanol dissolved proteins other than proteolipids, i.e., simple solubility in chloroform-methanol was not enough for a strict definition of proteolipids. In 1966, Wolfgram (41) described an acid-soluble proteolipid which was markedly different from the classical proteolipids, and subsequent authors traced the Wolfgram fraction to isolated myelin (7, 20).

Briefly, it appears at the present time that central myelin contains at least three different proteins or types of protein, to wit, the classical proteolipids, the basic antigenic protein responsible for allergic encephalomyelitis, and the Wolfgram fraction. The mutually confirmatory results of several groups (2,7,20) indicate that these three fractions represent approximately, proteolipids one-half, basic protein one-third, and Wolfgram's fraction one-sixth of central myelin proteins.

The proteins of peripheral myelin have received less attention than central myelin proteins, presumably because of the difficulties inherent in the isolation of peripheral myelin. However, it is clear that they are different quantitatively, and possibly qualitatively from the proteins from central myelin (7, 31, 42).

Up to now, the progress made in our knowledge of myelin proteins has thrown little light on the nature of neurokeratin, and its relationship, if any, with the three different fractions that are known to occur in central myelin. This may indicate an exclusively extramyelinic origin of neurokeratin. It is more likely, however, that neurokeratin (or TRPR) represents, at least in part, a degraded form of the classical proteolipids.

Fractionation of central myelin proteins. - The quantitative separation of the three different protein fractions from central myelin was carried out in 1968 by Eng et al. (7) by the use of 0.5 % Triton X-100 in 0.5 M ammonium acetate.

More recently the same result has been obtained by a simpler procedure by Gonzalez-Sastre in our laboratory (20). The method is based on the observations that Wolfgram's proteolipid is insoluble in neutral chloroform-methanol mixtures, and that the encephalitogenic basic protein is insoluble in chloroform-methanol in presence of electrolytes. In this method myelin, isolated by one of the standard methods of subcellular fractionation, is dissolved in chloroform-methanol. The solution is then centrifuged at about 1000 g for 10 min or until clear. A small amount of insoluble residue is thus collected. It amounts to 2 percent or 3 percent of the dry weight of myelin, i.e., 10 percent to 15 percent of the total proteins present. The supernatant is then diluted by addition of 1/20th its volume of 0.5 M KCl. A precipitate is formed which is collected by centrifugation. We thus have divided myelin among three different fractions : the original chloroform-methanol insoluble fraction (I), the subsequent fraction insoluble in chloroform-methanol-KCl mixture (II), and the final supernatant (III). These three fractions have been analyzed chemically, and by polyacrylamide gel electrophoresis. It has been found that I has the electrophoretic mobility and the amino acid composition of Wolfgram's proteolipid (Table I) ; that II has the mobility and the amino acid composition of the basic antigenic protein, and that III has mobility and an amino acid composition indistinguishable from those of classical proteolipids. The respective amounts of these three fractions as percent of total myelin protein are : Wolfgram's proteolipid : 15-17 percent ; basic protein: 30 percent ; classical proteolipids : 50-55 percent. These values are very close to those obtained by Eng *et al.* (7).

Proteolipids

In 1951, Folch and Lees (12) described the presence in brain tissue of a protein material which was insoluble in water and aqueous solvents and soluble in chloroform-methanol mixtures. Since all other known proteins were completely insoluble in these solvent mixtures, it was assumed that the proteins involved were combined with a lipid moiety that lend to the complex its particular lipid-like solubility. To designate this new type of substances, they coined the term proteolipid to indicate that they were lipoproteins with some of the physical properties of lipids, in sharp contrast to other then known lipoproteins which were soluble in water and aqueous solutions, and, not only insoluble in chloroform-methanol, but usually rendered insoluble in water ("denatured") by exposure to this particular solvent mixture.

The first indication of the occurrence of proteolipids was the observation that chloroform-methanol extracts from brain tissue, presumably freed of non-lipid substances by washing with water (11, 14), contained protein material, and that this material was not removed by water washing, even when the washing was repeated up to

TABLE I

Comparison of the Amino Acid Compositions of the Three Myelin Fractions with those of Wolfgram Proteolipid, the Basic Protein of White Matter and the Proteolipid Protein

Amino acid	CM (2:1,v/v) insoluble(+)	Wolfgram (1966) proteolipid	Basic protein from myelin (++)	Basic protein of white matter (Martensson and Le Baron, 1966)	Trypsin-resistant protein from myelin(++)	Proteolipid protein (Tenenbaum and Folch-Pi, 1966)
Lysine	6.87 ± 0.48	6.95	7.79 ± 0.49	7.75	4.12 ± 0.12	4.3
Histidine	2.06 ± 0.14	2.29	5.04 ± 0.23	5.67	2.02 ± 0.21	1.9
Arginine	6.14 ± 0.76	5.83	9.47 ± 0.57	10.14	2.50 ± 0.49	2.6
Aspartic acid	9.47 ± 0.35	9.90	6.71 ± 0.21	6.99	5.03 ± 0.06	4.2
Threonine	5.29 ± 0.10	5.16	3.71 ± 0.50	4.15	7.68 ± 0.69	8.5
Serine	6.00 ± 0.27	5.84	9.46 ± 0.55	9.71	5.84 ± 0.53	5.4
Glutamic acid	14.27 ± 1.08	12.95	7.63 ± 0.26	6.37	6.77 ± 0.06	6.0
Proline	5.00 ± 0.40	4.64	7.24 ± 0.40	7.44	3.03 ± 0.04	2.9
Glycine	7.16 ± 0.39	8.00	15.49 ± 0.40	15.08	11.91 ± 0.38	10.3
Alanine	8.96 ± 0.31	8.45	8.90 ± 0.61	8.81	12.28 ± 0.21	12.5
Half cystine	1.09 ± 0.17	1.01	0.00	0.00	4.22 ± 0.04	4.2
Valine	5.56 ± 0.10	5.84	1.38 ± 0.15	1.54	7.11 ± 0.44	6.9
Methionine	1.77 ± 0.26	2.13	1.28 ± 0.12	1.16	1.17 ± 0.22	1.7
Isoleucine	4.94 ± 0.22	4.27	1.57 ± 0.19	1.76	5.10 ± 0.49	4.9
Leucine	8.81 ± 0.15	9.63	6.02 ± 0.25	6.35	10.98 ± 0.81	11.1
Tyrosine	2.86 ± 0.13	2.87	2.79 ± 0.20	2.49	3.61 ± 0.40	4.7
Phenylalanine	3.78 ± 0.42	4.22	5.36 ± 0.32	4.78	6.62 ± 0.22	7.8

Data are expressed as moles/100 moles of total amino acids uncorrected for hydrolytic losses.
(+) Average of four preparations ; (++) Average of two preparations ; both ± S.D.

six times. The presence of this protein material in the extract was revealed by the fact that when the unwashed extract was taken to dryness by evaporation of the solvents, the resulting residue proved to be only partly soluble in chloroform-methanol, or in water. The fraction insoluble in both these solvents was found to be a protein that contained 14 % N, 1.76 % S, and which, upon adequate acid hydrolysis, liberated over 91 percent of its N as free amino acids (Table II). From the washed extract, the same material could be isolated by taking it to dryness provided the extract was first diluted with methanol and water to a final solvent composition similar to that of the unwashed extract, i.e., the washed extract was diluted with one-half its volume of methanol, and about 7 percent its volume of water. The presence of substantial water in the mixture, which resulted in the formation of a biphasic system in the course of the evaporation of the solvents, was a necessary requirement for the insolubilization of the protein material. Apparently, the protein in solution in the chloroform-methanol would accumulate at the interface of the biphasic system and there undergo a "denaturation" process which would render it generally insoluble.

TABLE II

Amino Acid Composition of Protein Moiety from White Matter Proteolipid

Constituent amino acids	Amino acid (g/100 g) protein
Leucine and-or isoleucine	23.5
Methionine	3.8
Tyrosine	10.6
Proline	2.3
Alanine and glutamic acid	14.4
Threonine	7.2
Aspartic acid	4.1
Serine	5.3
Glycine	6.1
Arginine	3.2
Lysine	4.8
Histidine	2.4
Cystine	3.8
Total	91.5

Distribution of proteolipids. - Although especially abundant in nervous tissue, proteolipids are also found in a wide variety of animal and vegetable tissues. Bovine tissues that have been analyzed for proteolipids contain the following amounts of proteolipid protein (mg/g fresh tissue) : heart muscle, 3.5 : kidney, 2.0 ; liver, 1.6 ; lung, 0.95 ; uterus, 0.6 ; biceps, 0.4. In spinach chloroplasts (44) they represent 2-4 percent of dry weight. These values are only indicative because the yields obtained may have been incomplete. In general, proteolipids appear to occur in membranous structures : in heart muscle they are recovered mainly in the mitochondrial fraction (21), in central white matter, they are mainly myelin components ,(2) and ,in gray matter,they are found in various subcellular membranes (23,25).

In the nervous system, proteolipids are found at highest concentration in white matter (20-27 mg/gram fresh tissue), and at about 1/5 to 1/10th this concentration in gray matter (1,12). In peripheral tissue they are present at only 1/20th to 1/80th their concentration in white matter (9, 13). They are absent, or present below the level of detection, in fetal brain and their appearance and progressive accumulation is concurrent with myelination (10).

Purification of central white matter proteolipids. - A large amount of work has gone not only into the purification of proteolipids proper, i.e., the separation of the protein-lipid complex from adventitious lipids, but also into the preparation of the protein moiety free of lipids, i.e., the proteolipid apoprotein. It has been found that proteolipids can be concentrated by solvent fractionation, by differential centrifugation, by dialysis in organic solvents, by gel permeation, or by a combination of these procedures. The specific procedures that have been developed are i) the "fluff" method of Folch and Lees (12) which was the one first used for the separation of proteolipid-enriched fractions from the tissue chloroform-methanol extract. The extract is overlaid with at least 5-fold its volume of water. Eventually, a "fluff" accumulates at the interface, which can be collected by freezing. From this fluff, by solvent fractionation, two preparations, proteolipids A and B are obtained. From the chloroformic solution underlying the fluff, a proteolipid C is obtained by solvent fractionation. ii) The emulsion-centrifugation procedure (15) is based on the difference between the specific gravity of protein and that of lipids. The lipid and proteolipid mixture recovered from a washed lipid extract (14) is emulsified in 30-fold its weight of water ; by centrifugation, proteolipids are collected quantitatively at the bottom (crude proteolipid), leaving the majority of lipids in emulsion in the supernatant. From the pellet, by solvent fractionation, a "concentrated proteolipid" preparation is obtained. iii) By dialysis in organic solvents (27, 33), free lipids diffuse through the dialysis membrane, whereas the proteolipids are retained by it. At the completion of dialysis, the retentate contains proteolipids freed of the greater part of lipids. iv) In gel permeation, as in dialysis, the proteolipids are separated from free lipids because of their larger molecular size. Using polystyrene gel, Autilio (2) was able to separate proteolipids from isolated myelin. Later, Mokrasch (32), by combining precipitation with ethyl ether with permeation on Sephadex LH20, obtained a highly enriched preparation of proteolipids. Finally, Soto et al. (35) used also Sephadex LH20 in the purification of proteolipids from white matter and from gray matter. From the latter they were able to separate a distinct fraction which exhibited many of the properties postulated for an acetylcholine receptor, and for which the function of physiological receptor is being claimed (5).

Since these procedures were not always designed to yield proteolipids of the highest possible protein content, the different products obtained show a wide range of variation in composition (Table III). The preparations with the highest protein content, which are those obtained by dialysis or by gel permeation, still contain about 15 percent lipids or more. These lipids, which are the most firmly bound to the protein, are mainly, if not exclusively, phosphatidylserine, sulfatides and polyphosphoinositides, i.e., they are acidic lipids. They appear to be bound to the protein moiety by ionic linkages. They can be removed in part by chromatography on silicic acid (30), or on a Dowex 1-X2 (32), both procedures yielding products with about 95 percent protein (Table III). To remove the lipids completely, however, it is necessary to submit the proteolipid to dialysis in chloroform-methanol acidified by addition of concentrated HCl to a final HCl concentration 0.04 N. A protein is then obtained which contains only traces of some lipids, with the exception of 2 to 4 percent covalently bound fatty acids (see below). It is the proteolipid apoprotein.

The chromatography of proteolipids on silicic acid has been carried out by Matsumoto et al. (30) with preparations obtained by the emulsion-centrifugation procedure. Typical conditions are a 10 mm inner diameter column packed with 4 g silicic acid and loaded with about 70 mg of a preparation containing 60 percent protein.The column is eluted with a discontinuous gradient of chloroform-methanol-water mixtures of increasing polarity starting at 85 percent chloroform and ending with chloroform-methanol (1:1, v/v) containing 12 percent 0.05 N HCl. Free lipids appear at 85 and 75 percent chloroform, and three protein peaks appear at 75, 70 and 50 percent chloroform (acidified). No difference is found among the three protein peaks in amino acid composition (Table IV) and in the nature of the up to 5 percent lipids that each contains, which are mainly polyphosphoinositides.

The preparation of maximally delipidated apoprotein was first carried out by Tenenbaum and Folch-Pi (37). It consisted in dialyzing a washed lipid extract against chloroform-methanol (2:1,v/v)

TABLE III

Average Composition of Brain White Matter Proteolipid and Apoprotein Preparations Obtained by Different Procedures

Unless otherwise noted, all components are expressed as % of the respective preparation

Procedure and preparation(s)†	Protein	P	Phospholipids P x 25	Neutral sugars	Cerebrosides & sulfatides	Cholesterol	$E^{278\ nm}_{1\ cm/1\%}$
Fluff method							
-Proteol. A	12-20	0.2-0.6	5-15	11.0	50	traces	-
B	35-40	1.2	30	4.4	20	traces	-
C	55-70	1.0-1.2	25-30	traces	traces	traces	-
Emulsion-centrifugation							
-Crude proteolipid	35-35	1.0-1.2	25-30	1.1-2.9	5-13	7	6-8
-Concentrated	55-60	1.0	25	0.6-1.2	3- 7	1	9-11
Silicic acid chromatography							
I	95	0.08-0.2	2-5	0.2	1	-	13-14
II	95	0.08	2	0.2	1	-	13-14
III	95	0.08	2	0.2	1	-	13-14
Sephadex LH20	75	0.68	16.5	2.34	10.6	-	13.2
+ Dowex 1X2	94	0.074	1.9	0.47	2.0	-	16.3
K_3 citrate denatured proteolipid	97-98	0.2	2.0††	-	-	-	-
Dialysis in CM (Lees *et al.*)	70-85	0.4-1.0	10-25	0.5-1.2	2.5	0.2	10.5-13.5
+CM.HCl (Tennenbaum and Folch-Pi)	99-100	0.04	?	0.1	?	-	16-22
Present apoprotein (Folch-Pi & Stoffyn)	99-100	0.01-0.04	?	0.1	?	-	13.6

†Yields: Proteolipids A, B, and C jointly account for all the tissue proteolipids. Emulsion-centrifugation fractions also account for all of the proteolipids. Silicic acid fractions jointly account for 75 % of proteolipid placed on column. Yield of K_3 citrate denatured proteolipid depends on conditions. Dialysis methods yield proteolipids quantitatively minus the dialyzable fraction. ††Phospholipid in this preparation is polyphosphoinositide. Hence factor used is 10 instead of 25.

TABLE IV
Amino Acid Composition of Brain White Matter Proteolipid Prepared by Different Methods
Results expressed as % of total α-amino acid N recovered from acid hydrolysates

Amino acids	E-C Concentrated proteolipid	Fraction from silicic acid chromatography			Present Apoprotein
		I	II	III	
Leucine	11.4	11.8	11.1	11.4	11.1
Isoleucine	4.9	5.3	4.9	4.8	4.9
Valine	6.7	6.7	6.5	6.4	6.9
Glycine	10.7	9.6	10.5	10.8	10.3
Threonine	8.3	9.0	9.0	8.8	8.5
Serine	(7.4)+	(5.5)+	(5.2)+	(5.0)+	(8.5)+
Proline	3.1	3.0	2.3	2.4	2.8
Aspartic acid	4.0	4.3	3.9	3.9	4.2
Glutamic acid	5.8	4.3	3.8	4.2	6.0
Histidine	1.9	1.7	1.8	2.0	1.8
Arginine	2.0	2.7	3.1	3.2	2.6
Lysine	3.8	3.9	4.3	4.4	4.3
Tyrosine	4.9	4.4	4.9	4.9	4.6
Phenylalanine	8.3	8.6	8.2	8.2	7.9
Alanine	12.0	13.3	13.5	13.2	12.5
Methionine	1.7	1.5	1.7	1.8	1.9
Half-cystine	4.4	4.5	4.0	4.7	4.0

+ Serine computed uniformly at 6 % of total amino acids.

(CM) for seven days with daily changes of the diffusate, followed by dialysis for seven more days against chloroform-methanol-HCl, 2:1:0.04 N (CM.HCl), followed by dialysis against a series of outer phases in which water was replacing gradually the organic solvents, until those were completely eliminated. The final retentate contained a "water-soluble proteolipid protein".

Properties and composition of proteolipids. - The different proteolipid preparations obtained by the foregoing procedures are freely soluble in chloroform-methanol mixtures, and insoluble in water and in aqueous solutions. In the biphasic system chloroform-methanol-water (8:4:3, v/v) they will concentrate quantitatively in the lower chloroformic phase, and will be completely absent from the upper methanol-water phase (with the exception of the last protein peak eluted from silicic acid column : about 1/5 of the protein partitions into the upper phase). The retention of the original solubility properties throughout the gradual removal of lipids forces the conclusion that the characteristic solubilities of proteolipids must be explained in terms of the conformation of the protein moiety. Studies of some of these preparations by optical rotatory dispersion (ORD) (34, 43) show that proteolipids are characterized by a high content of α-helix which remains unchanged throughout the procedure of gradual delipidation. Upon removal of the last traces of lipids, the resulting apoprotein proves to be soluble in water, although still retaining its solubility in organic solvents. The newly acquired solubility in water is paralleled by a decrease of the α-helix content to below the detectable limits (< 10 %), i.e., there is a conformational change which apparently makes available to the medium the hydrophilic groups of the protein which in the starting proteolipids must have been buried inside the molecular structure.

The solutions of proteolipids in chloroform-methanol (2:1,v/v) are very stable and they keep for years without developing turbidity or precipitates, even at room temperature. They can be taken to dryness by evaporation of the solvents without loss of solubility of the proteolipids in the residue, provided the evaporation takes place at 40° or lower, and provided no biphasic system develops in the course of the evaporation. The formation of a biphasic system results in partial or total insolubilization of the proteolipid protein unless corrected immediately by addition of a proper solvent (usually methanol). A similar result is obtained by exposing the proteolipid solution in a biphasic system to slight alkalinity, in presence of ions : for instance, at pH 8.8, in the biphasic system chloroform-methanol-water (8:4:3, v/v) proteolipids will become insoluble to an extent which is proportional to the logarithm of the ionic strength, between 0.001 and 1.0 M NaCl (39). The same result is obtained with Na_3 or K_3 citrate solution.

All the proteolipid preparations have been found to be resistant to the action of pepsin, trypsin, papain and erepsin. This resistance is not due to the presence of lipids because it persists in the apoprotein. The only enzyme that attacks proteolipids is pronase, although the extent of this susceptibility has not been thoroughly explored. In addition, Lees et al. (29) have reported that trypsin attacks proteolipids in presence of Triton X-100.

The chemical composition of the various proteolipids is given in Tables III and IV. Values given for proteolipids A, B and C are merely indicative because both yield and composition vary widely according to the exact conditions used. The other methods of preparation yield more consistent products. It can be seen that, whereas from the amino acid composition, the protein moiety appears to be the same in all preparations, the lipid "moiety" varies according to the procedure of preparation . Fractionation by solvents, emulsion-centrifugation, permeation on Sephadex LH20, and dialysis in CM separate, from the protein, cholesterol and the bulk of cerebrosides, and sulfatides and phospholipids; hence these lipids are bound to the protein, if at all, by labile bonds that are easily disrupted. Most remaining phospholipids are removed by chromatography on silicic acid leaving only polyphosphoinositides (30) ; the small amount of neutral sugars most likely correspond to sulfatides. With Dowex 1X2 a similar removal of lipids is attained, although no information is available on the exact nature of the lipids still remaining. Finally, with dialysis in CM.HCl, the highest degree of delipidation is obtained, which proves that the most tightly bound lipids must be bound through ionic linkages.

The amino acid composition of the different preparations up to and including the apoprotein, shows that the amino acid pattern of proteolipids is not affected by the previous procedures of purification. The amino acid pattern has the following characteristics : a) a relative scarcity of basic and acidic amino acids : arginine, lysine, and histidine account jointly for less than 10 percent of amino acids in the hydrolysate, and aspartic and glutamic amount jointly to about 10 percent. b) There is a relative abundance of the so-called non-polar amino acids, i.e., amino acids that, when combined in a peptide chain, offer only non-polar groups to the medium : leucine, isoleucine, valine, glycine, proline, phenylalanine, and alanine amount to 57-58 percent of amino acids. If tryptophan is added, about 60 percent of amino acids are non-polar. The relatively high concentration of tryptophan is indicated by the high absorption at 280 nm. c) There is a relative abundance of methionine and half-cystine, as is to be expected from the high concentration of sulfur in proteolipid protein (1.76 percent). Lees et al. (28), in a study of the conditions necessary for the preparation of the carboxymethylcysteine derivatives of proteolipids, have shown that the protein contains both sulfhydryl and disulfide groups, but the sulfhydryl groups are difficult to demonstrate. They are available for reaction with alkylating agents only in the presence of sodium dodecyl sulfate, and a portion of them react slowly. Approximately one-third of the half-cystine residues exist in the sulfhydryl form when exposed to SDS ; the remainder occurs in disulfide linkages which must be reduced before alkylation can occur.

Recent Studies on Proteolipids

Preparation of an improved preparation of proteolipid apoprotein. - The "water-soluble proteolipid protein" obtained by Tenenbaum and Folch-Pi (37) was in the form of a final aqueous retentate which contained about 0.1 percent protein. The protein precipitated at neutral pH. If lyophilized, the residue proved either insoluble in water or soluble only below pH 5, and even this solubility was lost upon standing a few days. The protein was still soluble in CM. ORD measurements showed it to have no measurable α-helix content, compared to the high helicity exhibited by the starting proteolipid.

Because of its instability, this preparation did not lend itself to study by the classical methods of protein chemistry. Therefore, the whole question of its preparation was reinvestigated with the result that an improved and simplified method of preparation has been developed. It yields a consistent and stable preparation of proteolipid apoprotein, upon which extensive studies have been carried out. The scrutiny of this procedure has dealt with (i) conditions of extraction, (ii) conditions of dialysis with a view to maximal efficiency of delipidation and minimal exposure to acid and (iii) study of the process of passage from CM to water medium.

(i) In the classical procedure of extraction (11) the tissue is homogenized with 19-fold its volume of chloroform-methanol (2:1, v/v) (CM), the homogenate filtered, and the filtrate treated by addition of 1/5 its volume of water. The lower phase from the resulting biphasic system represents essentially a total lipid extract, and contains the proteolipids. Only polyphosphoinositides remain in the CM insoluble residue ; they are extracted from it with C-M-12 N HCl (200:100:1, v/v) (CM.HCl).

The procedure has been modified and consists now of homogenizing one volume of tissue with five volumes of chloroform-methanol (1:1, v/v) and 0.50 volumes of aqueous 2M KCl. In the resulting biphasic system, the lower phase contains essentially all of the tissue lipids and proteolipids, including the polyphosphoinositides (PPI) (Hayashi *et al.*, personal communication). The upper phase contains gangliosides, non-proteolipid protein, and low molecular weight tissue components.

The new procedure represents a gain in manipulation and time, and a saving of solvents. It does not seem to affect the quality of the final product, and the apoprotein it yields appears to be identical in all aspects with the one obtained from extracts prepared by the classical procedure.

The extractibility of polyphosphoinositides under these conditions corresponds to the experience of Garbus et al. (19) who first described this technique for the extraction of polyphosphoinositides from mitochondrial suspensions.

(ii) Dialysis has been followed under varying conditions, both as to rates and as to the nature of diffusates and retentates. The main points established are : a) Dialysis proceeds fastest in CM-HCl (2:1:0.01, v/v). In neutral CM mixture, the rate of dialysis is affected adversely by the presence of substantial amounts of water, and, near saturation, (ca. 8 percent) dialysis ceases altogether. b) In neutral solvents, dialysis will yield eventually proteolipids containing up to 80-85 percent protein and 15 to 20 percent lipids. These are mainly acidic lipids and presumably combined with the protein through ionic linkages.

In the first stage of dialysis protein material diffuses through the membrane to a variable extent, usually 10 to 15 percent of the total protein present. It has been possible to recover this dialyzable protein and to analyze it (17). It has proved to be undistinguishable from classical proteolipids, in resistance to proteolytic enzymes, end-groups (40) and presence of fatty acids bound covalently to the protein (see below) ; its amino acid composition is also undistinguishable from that of classical proteolipids in all features except for a lower content of half-cystine in some of the preparations. It has not yet been established whether this difference is fortuitous or significant.

(iii) Dialysis in CM-HCl is a necessary step for completion of the delipidation of the proteolipid protein. Besides protonating acidic lipids, thus permitting their removal by dialysis, the exposure to acid may have a chemical effect : sulfatides are desulfated, and polyphosphoinositides and phosphatidylserine are deacylated in a matter of hours in the acidified solvent mixture that is used. The protein appears to have a protective effect against this chemical hydrolysis by the acid because proteolipid protein exposed to the acid medium for several days still contains sulfatides detectable on TLC. Hence, the mechanism of action of the acid may be the dissociation of the protein-acidic lipid ionic linkage by protonation, followed by dialysis of the resulting free lipid. The persistance of sulfatides in acid in presence of protein for a length of time several-fold that required for their hydrolysis in the absence of protein, would indicate that the dissociation of the lipids from the protein may require some time, possibly because some of the ionic lipid-protein linkages would only become available for protonation after some intramolecular rearrangement. Since complete delipidation can be obtained as well at -12° as at room temperature, it appears on the whole that the basic mechanism of delipidation by acid is protonation, and that chemical hydrolysis plays only an ancillary role.

After removal of acidic lipids, proteolipid protein may form complexes with sphingomyelin, upon return to a neutral medium. Whether due to some rearrangement in the protein molecule, or simply a consequence of the removal of the competing acidic lipids, these de novo formed sphingomyelin-protein complexes are not dissociated by additional dialysis in acidified chloroform-methanol, with the result that the final retentate consists of the sphingomyelin-protein complexes instead of the delipidated apoprotein. To prevent the formation of these complexes, dialysis in acid medium must be continued until 85 percent or more of the solutes in the starting extract have diffused out (Table V).

Protracted exposure to acid decreases the amount of fatty acids esterified in the apoprotein (Table V).

(iv) The passage of the retentate from chloroform-methanol to water by successive dialysis against mixtures of changing composition has been completely eliminated. It has been replaced by fast removal of the solvents in a stream of nitrogen by evaporation to dryness in a vacuum. In this procedure chloroform is removed mainly as an azeotrope with water.

Preparation of white matter proteolipid apoprotein. - Central white matter is extracted by the two phase-KCl method and the lower phase obtained by centrifugation is dialyzed against several changes of 10-fold its volume of CM, preferably in the dark and at 4°. When 2/3 of the starting total solutes have dialyzed out, dialysis is continued against CM.HCl, until at least 85 percent of the solutes in the starting lower phase have been removed. Then dialysis is continued further for at least 5 changes of neutral CM, or until the solutes in the outer fluid amount to 0.05 percent of less of the starting solutes, whichever is longer. The length of time required to reach these levels of dialysis varies from sample to sample of dialysis tubing and it can be as little as 2 days to reach diffusion of 2/3 of starting solutes, and 4 days to reach more than 85 percent, to two-fold or three-fold these lengths of time.

The final retentate is usually clear and contains about 5 mg apoprotein per ml. Upon evaporation, it yields the apoprotein as a residue which has a characteristic glass-like appearance in the

TABLE V

Effect of Different Conditions of Dialysis on the Properties of the Final Retentate (Apoprotein)

Exp.	% of the original solutes remaining in retentate upon passage from acid to neutral media	Exposure to acid		Properties of retentate at completion of dialysis			
		Temp.	Time	P %	$E^{280\ nm}_{1\%/1cm}$	Fatty acids %	Solubility in water
69-XXI	10.5	-12	15 days	0.037	13.6	3.6	soluble
69-XVII	12	22°	5 days	0.013	13.5	2.0	soluble
69-XIX	14	22°	3 days	0.019	13.5	2.0	soluble
70-III-2	15	-12	15 days	0.023	13.6	3.2	soluble
70-III-1	50	-12	15 days	0.32†	9.9	††	insoluble
70-VI-3	55	22°	3 days	0.26†	11.4	††	insoluble
70-IV	58	-12	15 days	0.21†	11.9	††	insoluble
70-VI-2	61	22°	2 days	0.18†	12.3	††	insoluble

†By TLC these retentates showed sphingomyelin as the main if not the only detectable phospholipid. Upon further exposure to acid media, there were only minor changes in P content and absorbance.
††Fatty acids not estimated because the presence of too much lipid material renders the values obtained worthless.

sense that apoprotein samples that have not been maximally delipidated will give a whitish residue.

Preparation of the water-soluble apoprotein. - The apoprotein recovered from the retentate is freely soluble in CM and in many other organic solvents,and it is completely insoluble in water. To render it water-soluble it is necessary to make it pass from solution into chloroform-methanol into solution in water by placing the CM solution in, or under, a stream of nitrogen, which removes chloroform preferentially (18). When about 4/5 of the weight of the solution has thus been removed, water is added to the concentrate until cloudiness develops, or until a volume of water equal to the concentrate has been added. Passage of nitrogen is continued until the weight of the solution has again been reduced by half ; the concentrate is again diluted with an equal volume of water, and the passage of nitrogen continued until the weight of the solution has once again been reduced by half. At this stage, the solution is essentially free of chloroform and methanol, and can be kept for further study as an aqueous solution of apoprotein. Alternatively, it can be taken to dryness in a vacuum desiccator. The residue, glass-like in appearance, will prove freely soluble in water and also in chloroform-methanol, although some time of contact between residue and solvent may be required and, in the case of chloroform-methanol, addition of 1 or 2 percent water to the mixture may be necessary to bring about complete solution. The residue retains these solubilities for as long as we have kept it, which is several weeks. Aqueous solutions up to 3 or 4 % can be easily prepared.Above 4 percent concentration, the aqueous solutions show increasing viscosity, and at 5 or 6 percent they become essentially gels.

It must be emphasized that only the maximally delipidated apoprotein is soluble in water. The presence of very small amounts of residual lipids result in incomplete insolubility in water. For instance, the sphingomyelin-protein de novo complexes (Table V) are quite insoluble in water.

The aqueous solutions of apoprotein are indefinitely stable at neutral or slightly acid pH's. They appear to be markedly, if not totally, resistant to bacterial contamination. In our experience, solutions stored at 4° but with frequent stays at room temperature, and also frequently opened for the taking of samples, have remained sterile for as long as they have been kept, which in some cases has been as long as eighteen months.

At pH 7.5 and above, aqueous solutions of apoprotein develop a turbidity which disappears at higher pH's, but persists upon acidification. Aqueous solutions brought rapidly to 0.1 N NaOH or higher alkali concentrations, develop a transient turbidity followed by complete clarification. This turbidity is so transient as to go unnoticed unless special attention is paid to its appearance and disappearance. The solutions in 0.1 N NaOH or higher, remain clear for several days at least but, upon acidification, there is formation of a massive precipitate which collects readily, leaving a supernatant that shows only negligible absorption at 280 nm, i.e., the apoprotein appears to have precipitated out quantitatively.This insolubilization of the apoprotein,exposed to 0.1 N NaOH,in aqueous acid develops over a period of about 2 hours.

The apoprotein is insolubilized, just as proteolipids are, by taking to dryness from biphasic solutions, and by exposure to low alkalinity in presence of ions (39). Thus, when a solution of apoprotein in chloroform-methanol (2:1, v/v) is diluted with one-fifth its volume of 0.1 M Na_3 citrate, the bulk, if not all, of the protein is precipitated out of solution, and collects as an insoluble residue at the interface. This residue is completely insoluble in all aqueous solutions and organic solvents that have been used, including 5 percent SDS (Na-dodecyl-sulfate). It has the same chemical composition as that of the apoprotein, including the amount of bound fatty acids. Treatment of the residue with CM.HCl fails to extract any of the bound fatty acids. Hence, it appears that exposure to alkali may render the apoprotein completely insoluble without any release of its bound fatty acids.

Although no systematic study has been made of the influence of salts on the solubility of the apoprotein in water, it has been observed that the apoprotein is precipitated out of solution by NaCl at about 0.27 M concentration. The precipitation is reversible,and upon elimination of NaCl, or lowering of its concentration, the apoprotein goes back into solution.

Physical properties of the apoprotein. - Apoprotein in aqueous solution exhibits one main peak and a much smaller second peak both in the ultracentrifuge and by moving boundary electrophoresis at both pH 7.0 and 5.0. In both analyses, the main peak accounts for 90 percent or more, and the second peak for 10 percent or less of the material in solution (Fig. 1, 2, 3, 4 and Table VI).

Under various conditions of polyacrylamide gel electrophoresis, several authors have reported single bands for proteolipid protein, using either isolated myelin (20), partly purified proteolipids (38) or apoprotein preparations (4). In our experience, proteolipid apoprotein gives one single, or one main band under various conditions of gel electrophoresis.

Briefly, it appears from physical evidence that proteolipid apoprotein is essentially a homogeneous protein. Whether the small second peaks observed in ultracentrifugation or in moving boundary electrophoresis represent contaminants or simply a different state of aggregation of the protein from the main peak cannot be decided without further study.

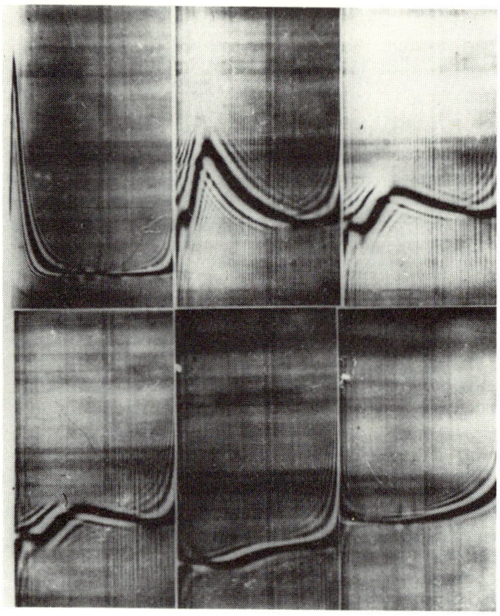

Fig. 1. Ultracentrifuge patterns of a 1.1 % solution of proteolipid apoprotein 68-XIV 22° H^+ in phosphate buffer, pH 7.0, $\frac{\lambda}{2}$= 0.1. Speed 24,000 rev/min. From left to right, top row, 0, 8, and 16 min running time. Bottom row, left to right, 20 and 30 min running time and then 36 min more at 52,000 rev/min.

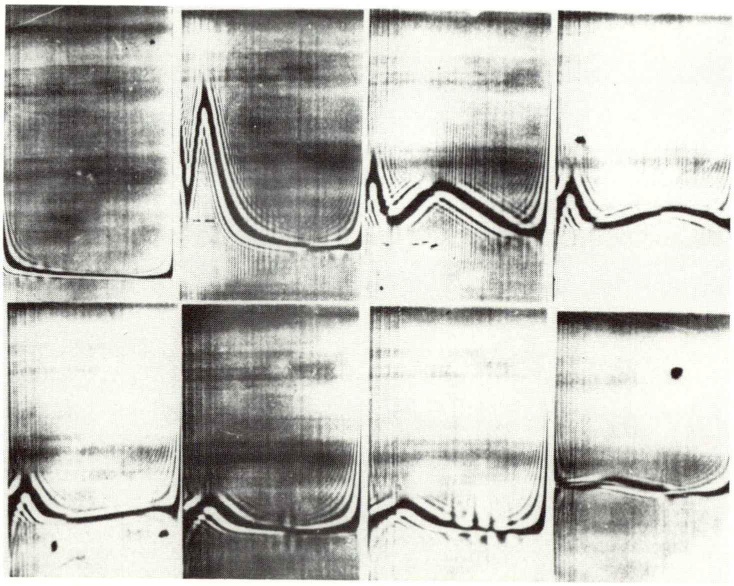

Fig. 2. Ultracentrifuge patterns of a 1.2 % solution of proteolipid apoprotein. Preparation 68-XIV 22° H^+ in Na-acetate buffer, pH 5.0 $\frac{r}{2}$ = 0.1 run at 26,000 rev/min. Top row, from left to right, 0, 8, 28 and 44 min. Bottom row, from left to right, 50, 92 and 124 min at 26,000 rev/min, and then 14 min more at 56,000 rev/min.

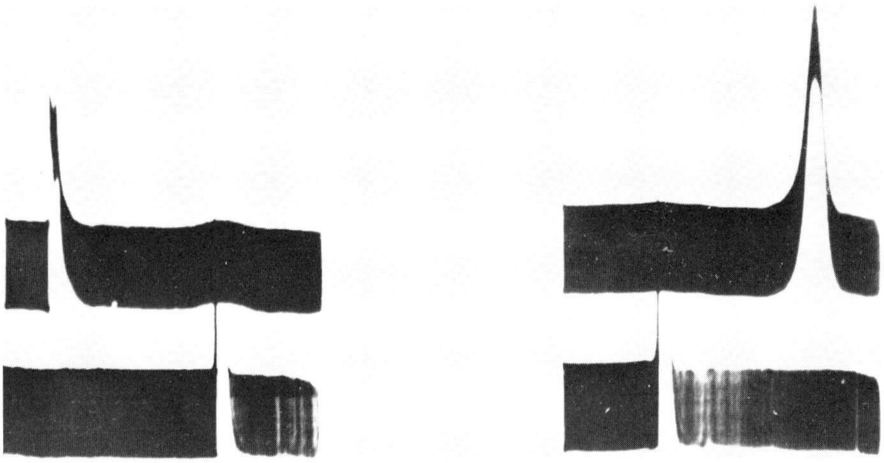

Fig. 3. Electrophoretic patterns of a 1.1 % solution of proteolipid apoprotein 68-XIV 22° H^+ in phosphate buffer, pH, 7.0 $\frac{r}{2}$ = 0.1.The ascending boundary is on the left, the descending boundary on the right. The scanning exposures below are initial boundaries; scanning exposures above, after 80 min of electrophoresis.

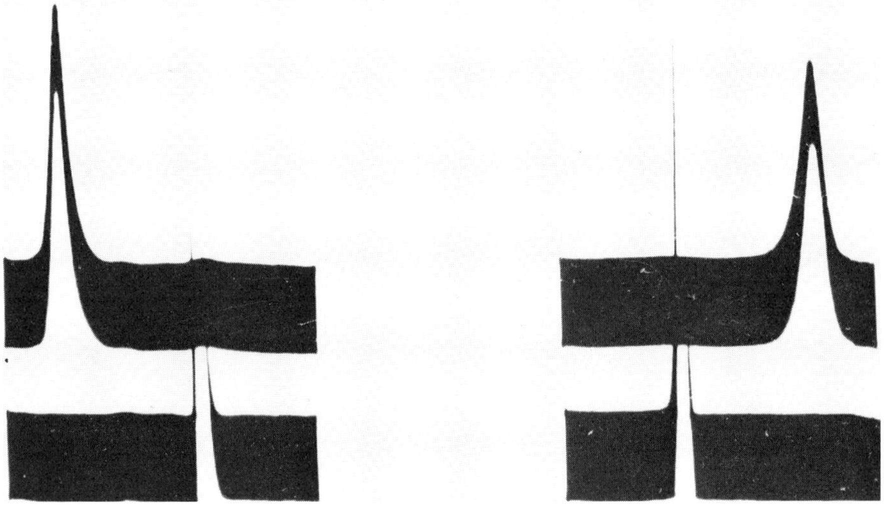

Fig. 4. Electrophoretic pattern of a 1.2 % solution of proteolipid apoprotein 68-XIV 22° H^+ in acetate buffer, pH 5.0 $\frac{r}{2}$ 0.1. The ascending boundary is on left ; the descending boundary is on the right. The scanning exposures below are initial boundaries ; scanning exposures above, after 44 min electrophoresis.

TABLE VI

Analysis of Proteolipid Apoprotein 68-XIV 22° H^+ by Electrophoresis and Ultracentrifugation

Solution	Free moving boundary electrophoresis				Sedimentation speed in Spinco Model E centrifuge			
	Component 1		Component 2		Peak 1		Peak 2	
	ux105(a)	Relative(b) %	ux105(a)	Relative(b) %	S	Relative %	S	Relative %
1.2% pH 5.0 Na Acetate $\frac{r}{2}$ = 0.1	+ 10.0	90 %	+ 8.4	10 %	8.2	5 - 10 %	62	90 %
						Speed : 26,000 rev/min(c)		
1.1% pH 7.0 phosphate $\frac{r}{2}$ = 0.1	+ 4.2	On ascending boundary limb resolves into two components: est.ratio 1:3 to 1:4			17	10 %	96	90 %
						Speed : 26,000 rev/min(c)		

(a) Electrophoretic mobility (cm^2/volt.sec.). (b) Relative concentrations are estimates.
(c) In both runs subsequent speeds of 52,000 and 56,000 respectively failed to show lighter component.

The apoprotein shows an absorption peak at 278 nm. Its $E^{1cm}_{1\%}$ at 278 nm is 13.6. This is lower than the values reported in other studies (Table III). The higher values may reflect contamination and, in some cases, the lack of correction for turbidity.

Relationship between the solubilities of the proteolipid apoprotein and its conformation. - The apoprotein occurs in two different forms, a lipophilic form soluble in organic solvents, and a hydrophilic form soluble in aqueous media. It can pass from one form to the other reversibly under the proper conditions of operation. To pass from the lipophilic to the hydrophilic form, it is necessary to follow the procedure described under "Preparation of water-soluble apoprotein". The reverse passage is much easier, and the apoprotein dried from aqueous solutions can be dissolved directly into CM, although the addition of a small amount of water may be necessary.

It is not known what exact changes occur with these reversible changes in solubility, but it is fair to assume that in the lipophilic apoprotein the lipophilic groups predominate at the surface of the molecule, whereas in the hydrophilic apoprotein, it is the hydrophilic groups that predominate at the molecular surface. Optical rotatory dispersion and circular dichroism measurements on solutions of apoprotein in different solvents (Table VII) bear out

TABLE VII

Optical Rotatory Dispersion Analysis of Typical Proteolipid Apoprotein (Results obtained on Sample 68-XIV-22A)

Solvent	Conc.(%)	% α-helix computed from		
		$b_\circ$*	$(m')_{233nm}$**	$a_\circ$
$CHCl_3$-MeOH	1.14	66	65	+ 131
$ClCH_2CH_2OH$	0.92	60	61	+ 97
F_3CCH_2OH	0.61	90	93	+ 69
H_2O	0.32	38	37	- 73
H_2O***	0.35	16	17	- 136
MeOH****	0.71	33	41	- 289
$CHCl_3$-MeOH ~60% lipid	0.81	69	66	+ 164

*$b_\circ$ = -630 for 100 % α-helix; $\lambda_\circ$ = 212 nm. **$(m')_{233nm}$ = -13,000 (deg.cm^2/decimole) for 100 % α-helix ; -1,700 (deg.cm^2/decimole)est. for 0 % α-helix.***Prepared in presence of excess MeOH.****When MeOH solution diluted with $CHCl_3$ to make 2:1, C-Me, the 233 trough shifts from 231 to 233.5, mμ, the helix becomes 33%. The MeOH seems to be a mixture of α-helix plus some other conformation that is not reversible.

that conformational changes occur, without defining their nature. Zand (43) reported originally that proteolipid was characterized by a high content of α-helix, whereas the "water-soluble" protein of Tenenbaum and Folch-Pi showed no measurable α-helix. Sherman and Folch-Pi (34) confirmed Zand's observations on proteolipids and extended them to show that the α-helix content did not change in the course of the gradual delipidation of the proteolipid to the apoprotein stage. They found, however, that in sharp contrast with the absence of measurable α-helix of the "water-soluble" apoprotein of Tenenbaum and Folch-Pi, the present apoprotein retained about one-half of the α-helix content of the CM soluble apoprotein, and that this change in conformation was reversible. Upon being placed back into solution in CM, the apoprotein regained its former α-helix content. This change appeared to be repeatedly reversible and a given sample of apoprotein could be changed back and forth repeatedly from its CM soluble form into its water soluble form with the corresponding changes in α-helix content.

Sherman and Folch-Pi (34) found that the reversibility of this change in conformation can only be preserved if exposure to water-methanol mixtures, or to pure methanol, is kept to a minimum. Apparently methanol changes the conformation of the apoprotein irreversibly. The result is an apoprotein with a reduced α-helix content which is soluble in methanol and which, if it goes into solution into CM at all, will do so without regaining its former α-helix content. In water, this apoprotein will form only very dilute solutions which gradually yield a precipitate which is insoluble in all the solvents that have been used. Presumably, it is this effect of methanol that is responsible for the instability and the low α-helix content of the water soluble apoprotein of Tenenbaum and Folch-Pi (37)

The contrast between the stability of solutions of apoprotein in organic solvents and in water, and the ease with which the apoprotein can be obtained as an almost universally insoluble residue (short of chemical breakdown) by such simple means as drying from biphasic solutions (12), exposing to low alkaline pH's in presence of ions (39) or exposure to methanol, suggests that the apoprotein is apt to undergo a number of changes in conformation, some reversible and some irreversible. None of these changes appear to change the chemical composition of the apoprotein.

Chemical composition of the proteolipid apoprotein. - The apoprotein does not show any spots on thin-layer chromatography for lipids even when samples as large as 5 to 10 mg are taken. It contains traces of P (0.01 to 0.04 %), and of carbohydrate (< 0.1 % as galactose). Its P is released by treatment with 0.1 N NaOH at room temperature for 16 h as water soluble organic P, i.e., it is not phosphoprotein P ; the water solution shows no absorption at 260 nm, which indicates that the P is not a nucleic acid derivative. As discussed below, this P most likely represents traces of phospholipids. The carbohydrate present is mainly galactose, and most likely corresponds to residual sulfatides and cerebrosides. In summary, the apoprotein is neither a phosphoprotein, a glycoprotein, nor a nucleoprotein.

The most striking feature of the chemistry of the apoprotein is the presence of from 2.0 to 3.2 percent fatty acids. These fatty acids (Table VIII) show a consistent pattern of about 60 percent palmitic, 25 percent oleic and 10 percent stearic acids with 5 percent other acids. Stoffyn and Folch-Pi (36) have established conclusively that these fatty acids are esterified since they do not react with diazomethane, which shows that they are not free acids, and they react with Na borohydride, with production of the corresponding alcohols, which shows that they are bound by ester linkages (Table IX).

These fatty acid residues do not belong to any recognizable lipid. An exhaustive analysis of the apoprotein for possible lipid moieties, with which these fatty acids might be bound, shows that the amounts of P, ethanolamine, choline, sphingosine, inositol, galactose and other sugars, sialic acid, and glycerol present in the apoprotein are individually and jointly unable to account for from less than one-tenth to no more than one-fourth, of the fatty acids present (Table X). In summary, it is necessary to conclude that these fatty acids esterified in the apoprotein, must be esterified on the polypeptide chain itself, except for the remote possibility that they are constituents of an as yet unidentified lipid. Such a hypothetical lipid would be singularly devoid of the most common moieties of lipids known at the present time.

TABLE VIII

Fatty Acids Combined in Proteolipid Apoprotein (PLA)

PLA preparation	Total fatty acids as % of weight of PLA (a)	Composition of fatty acid mixture as % of values in column a			
		Palmitic	Stearic	Oleic	Other
69-XVII	2.0	56	9.7	27	7.3
69-XIX	3.2	62	9.1	23	5.9
69-XXI	3.2	62	8.6	26	3.4
70-III-2	3.0	58.5	10.6	25.4	5.6
70-XII	2.45	60.0	9.1	26.0	4.9

TABLE IX

Fatty Acids in Intact Proteolipid Apoprotein (PLA) and after Various Lengths of Time of Hydrolysis by 0.2 N NaOH at 25° as Estimated by Methanolysis and by Reaction with Diazomethane ($CH_2N_2 + R.COOH \longrightarrow R.COOCH_3 + N_2$)

PLA Preparation 70-XII	Fatty acids as % of weight of PLA		Composition of fatty acid mixture as % of values in columns a or b			
	By methanolysis (a)	By CH_2N_2 (b)	Palmitic	Stearic	Oleic	Other
	2.45	none	58.5	10.6	25.4	5.6
After hydrolysis by 0.2 N NaOH at 25°						
30 seconds	2.45	traces	traces	traces	traces	traces
30 minutes	2.45	0.245	69.0	6.0	32.0	traces
4 hours	2.45	0.93	73	8	10	traces
72 hours	2.45	1.91	75	10	10	traces

The protein residue from the 72 h hydrolysis was acidified and extracted with Et_2O. It contained 0.40 % total fatty acids.

In other experiments, hydrolysates were acidified and extracted with Et_2O. The ether extracts gave the same fatty acid content by methanolysis and by CH_2N_2, indicating that fatty acids had been released as free acids.

TABLE X

Amounts of Possible Lipid Moieties Present in Proteolipid Apoprotein (PLA)

Fatty acids	2.2 to 3.0 %	or	83 to 113	micromoles per gram PLA
P :	0.01 - 0.03%	or	3 - 10	- - - -
Glycerol	< 0.03%	or	< 3	- - - -
Total carbohydrate including hexosamines	< 0.1 %	or	< 5.5	- - - -
Ethanolamine	< 0.02%	or	< 3.3	- - - -
Sialic acid	0.02 - 0.08%	or	0.7 - 2.4	- - - -
Choline, sphingosine, and inositol below the level of detection				

If it is assumed that all the P and all the carbohydrate present are lipid in nature, they would account for all the lipid moieties present other than fatty acids. They also would account for 7 to 25 micromoles of fatty acids. Hence the great majority of fatty acids present cannot be accounted for in terms of any known lipid.

The apoprotein exhibits an amino acid composition undistinguishable from that of the crude, and of the partly purified proteolipids (Table IV).

The apoprotein preserves intact the resistance to most proteolytic enzymes, that is characteristic of the proteolipids. On the basis of the least abundant residue being 2 methionines per mole of protein, the apoprotein appears to have 125 residues, giving a possible molecular weight of about 12,000 daltons (16).

The apoprotein dried to constant weight in high vacuum in presence of NaOH, gives aqueous solutions with pH's around 3.5. Apparently by the method described, the apoprotein is obtained as a fully, or almost fully, protonated anion. Titration of two different apoprotein preparations between pH 3.5 and 7.17 requires one μmole of NaOH for each 3.58 mg of apoprotein. Braun and Radin (4) report the use of one μmole of HCl for each 2 mg of apoprotein between pH's 6.0 and 3.0. Although the two sets of values are not strictly comparable, it is obvious that the apoprotein of Braun and Radin exhibited a larger number of titratable acid groups than do our preparations. A possible explanation for this discrepancy is that the procedure followed by Braun and Radin involved a much more prolonged exposure to acid than our own procedure, with the result that some glutamine residues may have been deamidated to glutamic acid.

The determination of end groups of the apoprotein has proved to be very difficult because of unforeseen difficulties inherent in the nature of proteolipids. With these limitations, Takahashi and Mokrasch (unpublished,personal communication), both by Sanger's method and by Edman degradation concluded tentatively that the amino end group was glycine. In a subsequent rigorous study, Whikehart and Lees by dansylation (40) have proved that the amino end groups are glycine and glutamic acid in the ratio of 3 to 1.

The molecular weight of the proteolipid apoprotein. - Although the coincident results of polyacrylamide gel electrophoresis, ultracentrifugation, and moving boundary electrophoresis, show that the apoprotein occurs in physically homogeneous forms, there is not the same agreement as to results that bear on its molecular size. In 1961, Folch-Pi computed from the amino acid composition a possible monomeric molecular size of about 12,500 (16). In 1969, Thorun and Mehl (38) using gel electrophoresis on an acrylamide density gradient, obtained a value of 34-36,000 for the molecular weight of their proteolipid preparation. More recently, Eng (6) using a 10 % acrylamide gel and phosphate buffer in 0.1 % SDS, has reported a molecular weight of 22,000-23,000 for his preparation. In addition, Folch-Pi has reported that a portion of proteolipid is dialyzable through cellophane membranes (17), presumable impermeable to molecules above a size of 12,000 daltons. This dialyzable fraction accounts for a greater or smaller fraction of the total proteolipid fraction according to the relative permeability of the cellulose membrane used. Paradoxically, the more permeable the membrane, the smaller the fraction of proteolipid protein that diffuses through it. In addition, this diffusion of proteolipid protein occurs only at the very beginning of the dialysis process, and it ceases completely, so that a second dialysate does not contain any protein. This behaviour suggests that in the original tissue extract, a major or minor fraction of the total proteolipid may be maintained in a diffusable monomeric form of dispersion by a factor or factors.This factor or factors is changed or eliminated during the dialysis process, and, with its removal, the proteolipid changes to an aggregate, undialyzable form.

Although it is impossible with the available evidence to establish with certainty the molecular size of the apoprotein, the facts already know are compatible with a monomeric size of around 12,000 (or one-half as much, if it is assumed that there is only one methionine residue per molecule). The values reported by Eng, and by Thorn and Mehl, would then correspond to the weights of a dimer and trimer, respectively.

Organization of the proteolipid molecule *in vivo*. - Although it is impossible to deduce with certainty, with our present knowledge, the actual organization of the proteolipid protein *in vivo*, some of the facts already know provide a frame of reference into which this molecular organization must fit. The most important is that since the isolated apoprotein is freely soluble in water, and the tissue proteolipids are completely insoluble in water, it is necessary to postulate that, in the tissue, the apoprotein must necessarily be combined with lipids. These lipids would stabilize the apoprotein in its lipophilic water insoluble conformation. The most likely candidates for the lipids most closely bound with the apoprotein are the acidic lipids, which would be combined with the cationic groups. This protein-lipid core might then be combined with other lipids by non-polar bonds.

ACKNOWLEDGEMENTS

The author acknowledges the support of part of this work by the U.S.Public Health Service Grants NS 00130 and FR 05484.

REFERENCES

1 AMADUCCI, L., J. Neurochem., 9 (1962) 153.
2 AUTILIO, L., Feder. Proc., 25 (1966) 764.
3 AUTILIO, L.A., NORTON, W.T. and TERRY, R.D., J. Neurochem., 11 (1964) 17.

4 BRAUN, P.E. and RADIN, N.S., Biochemistry, 8 (1969) 4310.
5 DE ROBERTIS, E., FISZER, S. and SOTO, E.F., Science, 158 (1967) 928.
6 ENG, L.F., Feder. Proc., 30 (1971) Abstr. 1248.
7 ENG, L.F., CHAO, F.C., GERSTL, B., PRATT, D. and TAVASTSTJERNA, M.G., Biochemistry, 7 (1968) 4455.
8 EWALD, A. and KÜHNE, W., Verhandl. Naturhist.-Med., 1 (1877)457.
9 FINEAN, J.B., HAWTHORNE, J.N. and PATTERSON, J.D.E., J.Neurochem., 1 (1957) 193.
10 FOLCH, J., In Biochemistry of the Developing Nervous System (ed. H.Waelsch), 1955, p.121.
11 FOLCH, J., ASCOLI, I., LEES, M., MEATH, J.A. and LEBARON, F.N., J. Biol. Chem., 191 (1951) 833.
12 FOLCH, J. and LEES, M., J. Biol. Chem., 191 (1951) 807.
13 FOLCH, J., LEES, M. and CARR, S., Exptl. Cell Res. Suppl.5 (1958) 58.
14 FOLCH, J., LEES, M. and SLOANE-STANLEY, G.H., J. Biol. Chem., 226 (1957) 497.
15 FOLCH, J., WEBSTER, G.R. and LEES, M., Feder.Proc., 18 (1959)228.
16 FOLCH-PI, J., Exptl. Ann. Biochim. Med., 21 (1959) 81.
17 FOLCH-PI, J., 3rd Intern. Meet. Intern. Soc. Neurochem., Budapest, 1971 ; Publ. Akademic Kiado, Abstr. p.239.
18 FOLCH-PI, J. and SHERMAN, G., 2nd. Intern. Meet. Intern. Soc. Neurochem., Milano 1969 ; (Ed. R.Paoletti, R. Fumagalli and C. Galli) Publ. Tamburini, Milano, p.169.
19 Garbus, J., DE LUCA, H.F., LOOMANS, M.E. and STRONG, F.M., J. Biol. Chem., 238 (1963) 59.
20 GONZALEZ-SASTRE, F., J. Neurochem., 17 (1970) 1049.
21 JOEL, C.D., KARNOVSKY, M.L., BALL, E.G. and COOPER, O., J.Biol. Chem., 233 (1958) 1565.
22 LAATSCH, R.H., Feder. Proc., 22 (1963) 316.
23 LAPETINA, E.G., SOTO, E.F. and DE ROBERTIS, E., J. Neurochem., 15 (1968) 437.
24 LEBARON, F.N. and FOLCH-PI, J., J. Neurochem., 1 (1956) 101.
25 LEES, M., J. Neurochem., 13 (1966) 1407.
26 LEES, M., J. Neurochem., 15 (1968) 153.
27 LEES, M.B., CARR, S. and FOLCH, J., Biochim. Biophys. Acta, 84 (1964) 464.
28 LEES, M.B., LESTON, J.A. and MARFEY, P., J. Neurochem., 16 (1969) 1025.
29 LEES, M.B., MESSINGER, B.F. and BURNHAM, J.D., Biochem. Biophys. Res. Commun., 28 (1967) 185.
30 MATSUMOTO, M., MATSUMOTO, R. and FOLCH-PI, J.,J. Neurochem., 11 (1964) 829.
31 MEHL, E. and WOLFGRAM, F., J. Neurochem., 16 (1969) 1091.
32 MOKRASCH, L.C., Life Sciences, 6 (1967) 1905.
33 MURAKAMI, M., SEKINE, H. and FUNAHASHI, S., J. Biochem., 51 (1962) 431.
34 SHERMAN, G. and FOLCH-PI, J., J. Neurochem., 17 (1970) 597.

35 SOTO, E.F., PASQUINI, J.M., PLACIDO, R. and LA TORRE, J.L., J. Chromatog., 41 (1969) 400.
36 STOFFYN, P. and FOLCH-PI, J., Biochim. Biophys. Res. Commun., 44 (1971) 157.
37 TENENBAUM, D. and FOLCH-PI, J., Biochim. Biophys. Acta, 115 (1966) 141.
38 THORUN, W. and MEHL, E., Biochim. Biophys. Acta, 160 (1968) 132.
39 WEBSTER, G.R. and FOLCH, J., Biochim. Biophys. Acta, 49 (1961) 399.
40 WHIKEHART, D.H. and LEES, M.B., Feder. Proc., 30 (1971) 1247.
41 WOLFGRAM, F., J. Neurochem., 13 (1966) 461.
42 WOLFGRAM, F. and KOTORII, K., J. Neurochem., 15 (1968) 1291.
43 ZAND, R., Biopolymers, 6 (1968) 939.
44 ZILL, L.P. and HARMON, E.A., Biochim. Biophys. Acta, 53 (1961) 579.

MYELIN BASIC PROTEINS

Marian W. KIES, Russell E. MARTENSON and Gladys E. DEIBLER
Section on Myelin Chemistry, Laboratory of Cerebral Metabolism, National Institute of Mental Health - Bethesda, Maryland 20014 (USA)

Our attention was first directed to a study of myelin proteins because of our interest in the induction of experimental allergic encephalomyelitis (EAE) and the fact that Kabat _et al._ (11) and Morgan (26) had reported that white matter was more encephalitogenic than gray matter. At that time the existence in myelin of a highly basic protein resembling certain of the histones was not even suspected. Early studies on the isolation of the encephalitogenic component of guinea pig central nervous system (CNS) tissue and its subsequent purification depended largely on bioassay of fractions for encephalitogenic activity (13). Although the basic nature and high cathodic electrophoretic mobility of the encephalitogenic protein were soon recognized (14), the available electrophoretic techniques were not adequate for critical monitoring of the purification procedures. Since the concentration of this protein in CNS tissue is high, its isolation was relatively simple, and sufficient quantities could be prepared for detailed studies of its chemical properties and structure.

The work of Laatsch _et al._ (5) established that the encephalitogenic protein is localized in myelin and clearly differentiated this myelin protein from basic proteins (histones) isolated from CNS nuclei. More importantly, this work provided us with a preparation of myelin basic protein (BP) which could be used as a reference in our studies of its isolation from whole tissue.

The techniques currently in use in most laboratories today are developments of those first developed in our laboratory (12) and in the laboratory of Dr. Roboz-Einstein (28) ; lipid is removed from whole CNS tissue by the Folch extraction technique (9) ; extraneous water-soluble proteins are removed by treatment of the tissue residue with water ; and BP is obtained in relatively high concentration and good yield by treatment of this insoluble residue with dilute acid. Even though the acid extract contains many other proteins, their concentration relative to that of BP is usually sufficiently small that the specific encephalitogenic activity of the extract (its ability to induce EAE in guinea pigs related to its protein content) is almost as high as the specific activity of purified BP.

We have found from our studies on guinea pig and rat BPs that the ease with which BP can be purified is influenced to a large extent by the pH at which the protein is extracted (17, 24). Extraction at pH 3.0 removes up to 90 percent of the total BP (24) but relatively little of the tissue histones. Only the very lysine-rich histone is extracted to an extent that may be significant,and this histone is easily removed from BP by any of several procedures (gel filtration, ion-exchange chromatography, solution in 5 percent TCA). BP is readily distinguished from most of the other proteins which are extracted from CNS tissue at pH 3.0 by its high cathodic mobility upon electrophoresis in acrylamide gels at pH 2.4 and its characteristic color when stained with amido black. In the electrophoretograms it appears as an intense blue-green disc (λ max 680nm) in contrast to the other protein discs which are grayish-blue in color (λ max 580 nm). No comparable electrophoretic component is present in liver extracts made under identical conditions. By utilizing these electrophoretic and staining characteristics much information can be obtained about BP from various sources simply by electrophoretic analyses of the pH 3.0 extracts of delipidated CNS tissue preparations without subjecting the BP to further purification (20).

We have utilized this technique to examine the prevalence of two BPs (rather than one) among various rodents (21). Myelin of the guinea pig resembles that of human, beef, rabbit, and monkey in that it contains a single BP. Rat myelin, in contrast, contains two BPs, the larger one (L) resembling the single BP of the other five species, and a second (S) which is somewhat smaller (18). The occurrence of two BPs has also been observed by other investigators in electrophoretograms of rat myelin proteins (1,3,6,25,34,35) and mouse myelin proteins (10, H.C.Agrawal, personal communication).In Fig. 1 are shown densitometric tracings of the electrophoretograms of pH 3.0 extracts of the CNS tissue of several rodents. These include rat, mouse, and hamster (suborder Myomorpha) ; guinea pig and chinchilla (suborder Histricomorpha) ; and woodchuck (marmot), prairie dog (prairie marmot) and squirrel (suborder Sciuromorpha). The characteristic doublet of myelin basic proteins of different size (L and S) occurs in representatives of two of the three suborders of rodents. If these few rodent species are truly representative of the three suborders, the occurrence of two BPs in mammalian myelin is quite widespread since 40 percent of all species of

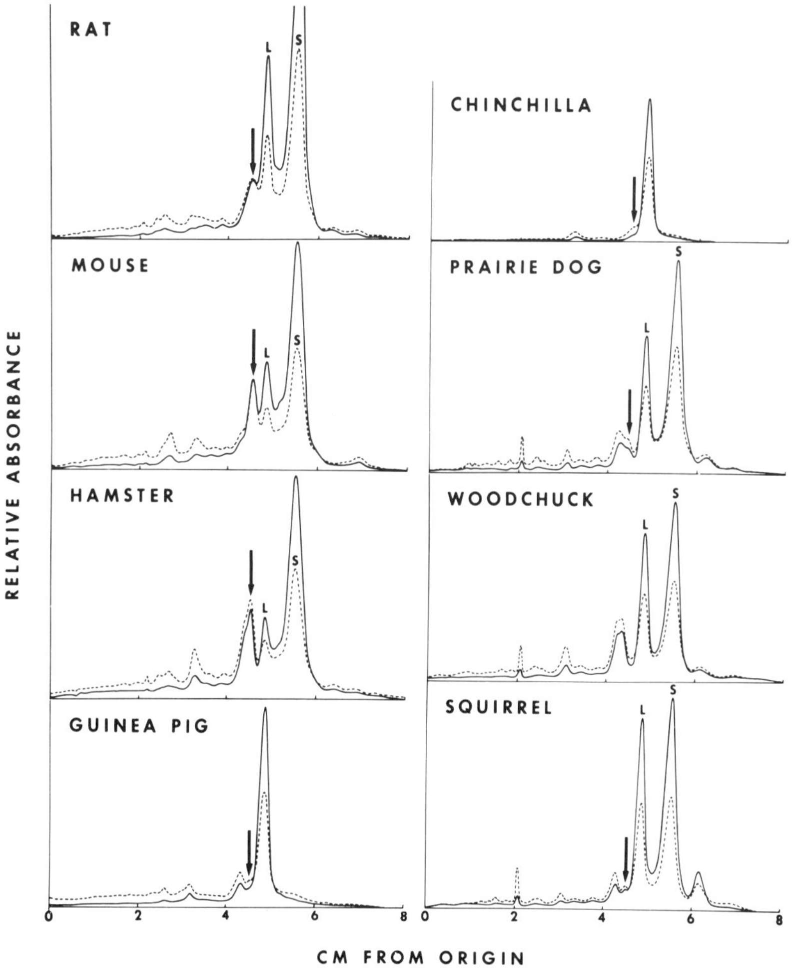

Fig. 1. Parallel electrophoresis at acid pH (1 M acetic acid, 8 M urea) in 5 % polyacrylamide gels of proteins (15-75 µg) extracted at pH 3.0 from rodent CNS tissues. Electrophoresis was carried out toward the cathode for 2.0 h at 2.5 mA per gel, proteins were stained with amido black, and gels were scanned at 690 nm (———) and 580 nm (-----). Rat, mouse, hamster, and guinea pig proteins were from brain ; the remainder were from total CNS tissue (brain plus spinal cord). The arrow points to histone F1 (very lysine-rich).The component located between the L and S BPs of rat, mouse, and hamster is probably a proteolytic breakdown product of the L BP.

mammals are rodents and 90 percent of the latter are in the two suborders Myomorpha and Sciuromorpha (27). In all of the species examined the S protein predominates. It is evident, therefore,that the shorter polypeptide chain possesses all of the information necessary for it to perform its structural function in myelin. These findings also suggest that the genetic events leading to the production of the additional smaller BP, presumably gene duplication followed by partial deletion, probably occurred only after the line leading to the suborder Hystricomorpha had been established and before the divergence of the remaining rodent line into the two suborders, Myomorpha and Sciuromorpha.

In addition to the several species already listed (guinea pig, human, bovine, rabbit, monkey, and the eight rodents) we have also examined BPs of several submammalian vertebrates : chicken, turtle (Chelydra serpentina), frog (Rana pipiens), shark (Carcharias obscurus), and carp. Chicken, turtle, and frog BPs are nearly identical electrophoretically to guinea pig BP at low pH. Shark BP has a slightly higher mobility than guinea pig BP, intermediate between the L and S rat proteins, while carp BP has a higher mobility at low pH than any of the others examined (Fig. 2). Our findings of

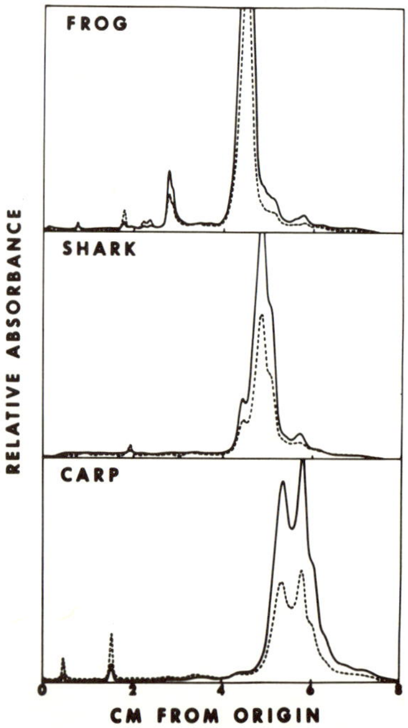

Fig. 2. Parallel electrophoresis at acid pH (1 M acetic acid, 8 M urea) in 5 % polyacrylamide gels of proteins (50 μg) extracted at pH 3.0 from spinal cord of some submammalian vertebrates.Electrophoresis was carried out, proteins were stained, and gels were scanned as described in Fig. 1.

identical electrophoretic mobilities for the chicken and guinea pig BPs at low pH compares with that of Mehl and Halaris (25). In contrast, however, we did not observe in the carp a second basic protein of relatively low mobility (25). The preparations of shark and carp BPs which we have examined have consisted of more than one electrophoretic component at low pH. Because the conditions of isolation did not preclude possible autolysis, we suspect that all but one of these components may be breakdown products of the native BP.

Many BPs which display essentially identical electrophoretic behaviour at low pH have been found to differ markedly in their electrophoretic behavior at high pH. It was first demonstrated by Martenson and Gaitonde (24) that bovine BP, while apparently homogeneous with respect to size, is heterogeneous with respect to charge. The single component observed in low pH electrophoretic patterns is resolved into several components on electrophoresis at high pH. As we have shown (19), the different components exist as such in the BP preparations and are not formed during electrophoresis. The different components can be separated by ion-exchange chromatography, but not by gel filtration, and each retains its characteristic electrophoretic behavior alone or in a mixture reconstituted to simulate the original preparations (4). At the present time, the explanation for these differences in charge is not entirely clear. There are several possible explanations which Martenson _et al._ (19) have discussed in an earlier report.

All of the BPs examined which appear to be homogeneous at low pH have been found to be heterogeneous by electrophoresis at a pH close to their isoelectric points. In those cases in which there is a mixture of two BPs of different size, e.g., in most rodents, the high pH electrophoretic patterns are even more complex because each of the two BPs contributes its own set of differently charged components. The high pH electrophoretic patterns of certain of the BPs (Fig. 3) reveal species differences in mobility which are not apparent when electrophoresis is carried out at acid pH. Electrophoresis at high pH also serves in some instances to "magnify" differences observed at low pH as shown by comparison of the mobility of the carp protein with those of some of the other species (Fig.4).

In order to carry out more detailed studies of the chemistry of the submammalian BPs, these proteins as well as those of rat, guinea pig, beef, and human were purified by gel filtration on Sephadex G-150, and in some instances by an additional step involving ion-exchange chromatography on carboxymethylcellulose. The molecular weights of all of the BPs examined, judging from the percent of total column volume at which they are eluted from the Sephadex columns, are the same (~ 18,400) except for the rat S and carp proteins. The carp protein appears to have roughly the same molecular weight as the rat S protein, ~ 14,400. Representative Sephadex G-150 elution patterns are illustrated in Fig. 5.

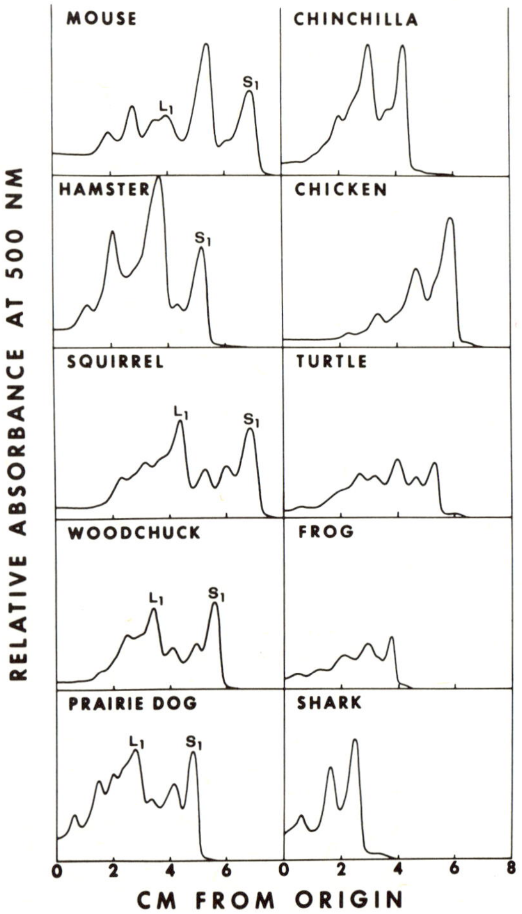

Fig. 3. Electrophoresis at alkaline pH (0.01 M sodium glycinate, pH 10.6, 8 M urea) in 5 % polyacrylamide gels of proteins (100-400 μg) extracted at pH 3.0 from CNS tissues. Electrophoresis was carried out toward the cathode for 3.0 h at 3.75 mA per gel. Under the electrophoretic conditions employed, BPs are essentially the only proteins in the extracts which migrate well into the gel. Although not all preparations were run in parallel, the differences in mobility are accurately represented. S_1 and L_1: most basic components of rodent S and L proteins, respectively. Rat and guinea pig preparations, respectively, were virtually identical with those shown for the mouse and chinchilla.

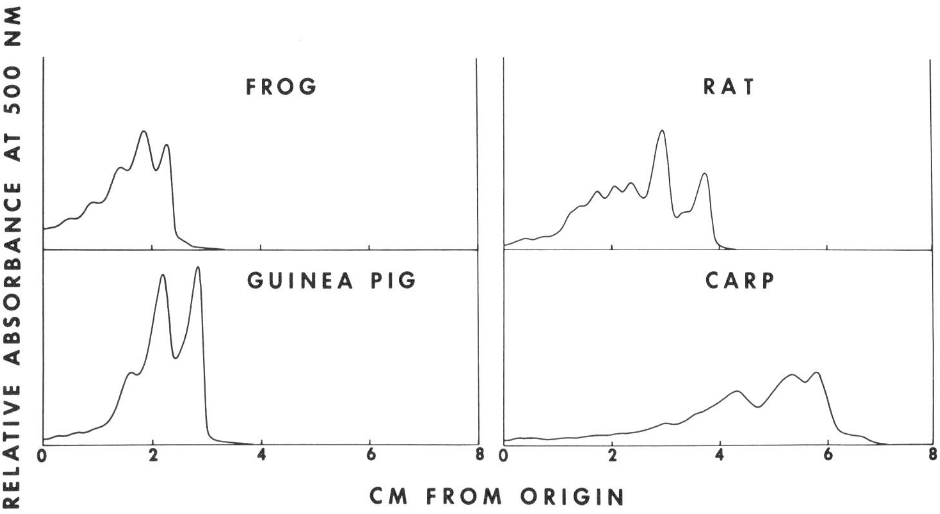

Fig. 4. Parallel electrophoresis at alkaline pH (0.01 M sodium glycinate, pH 10.6, 8 M urea) in 5 % polyacrylamide gels of proteins (100 µg) extracted at pH 3.0 from spinal cord. Electrophoresis was carried out toward the cathode for 1.5 h at 3.75 mA per gel.

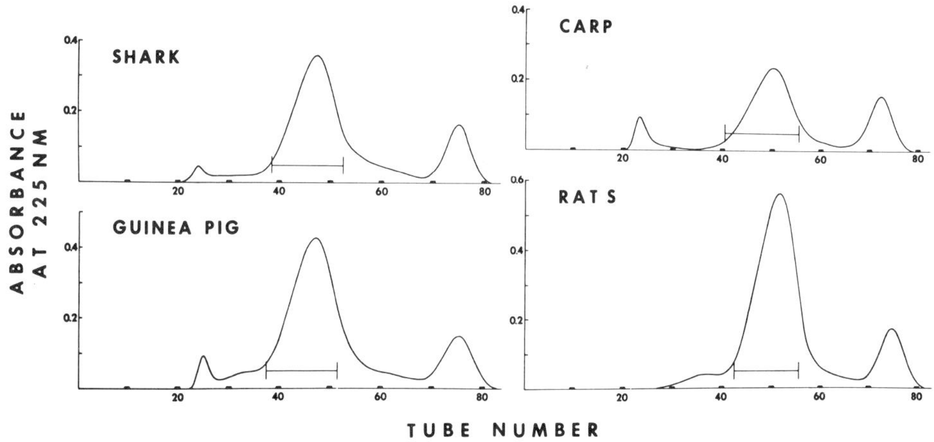

Fig. 5. Gel filtration of proteins extracted at pH 3.0 from spinal cord on columns (1.5 x 95 cm) of Sephadex G-150 equilibrated with 0.01 M HCl at 5°. Proteins (6.9-11.0 mg) were applied in 1.0 ml of 1 M acetic acid-0.01 M HCl, and fractions of 2.5 ml were collected at a flow rate of 15 ml per h and pooled as indicated. The last peak is acetic acid, which yields the total column volume (V_t). Guinea pig and shark BPs were eluted at 62 % of V_t ; rat S and carp BPs were eluted at 69 % of V_t.

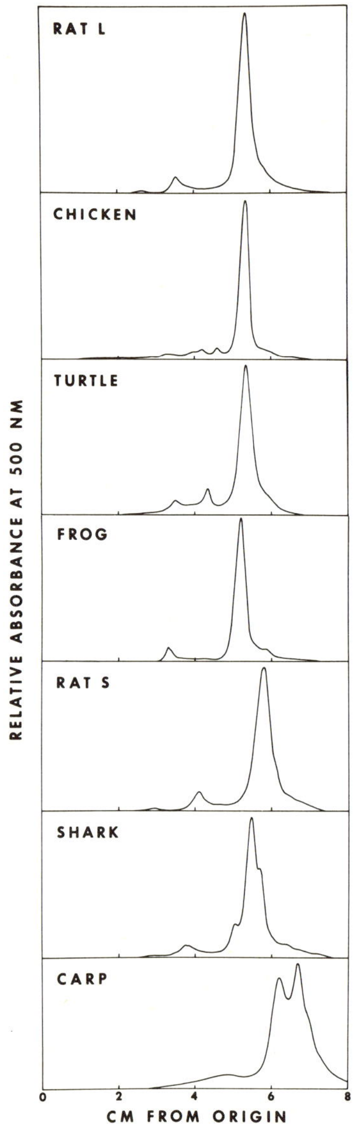

Fig. 6. Parallel electrophoresis at acid pH (1M acetic acid, 8M urea) in 5 % polyacrylamide gels of BPs (50 µg) purified by gel filtration on Sephadex G-150. Electrophoresis was carried out toward the cathode for 2 h at 2.5 mA per gel. The minor component of relatively low mobility in the rat and frog preparations appears to be a polymer of BP. Minor components of relatively low mobility in the chicken and turtle preparations are contaminants which were subsequently removed by ion-exchange chromatography.

Electrophoretograms of some of the purified BPs are presented in Fig. 6.

The reported amino acid sequences of bovine (7) and human (2) BPs are very similar. Guinea pig, rat L, chicken, turtle, frog, and shark BPs, which have not been sequenced, have amino acid compositions which, on the whole, are not too different from those of human and bovine BPs (Table I). All of these BPs contain 1 mole of tryptophan per mole of protein and all but the shark BP contain 2 moles of methionine per mole of protein. The tyrosine contents range from 2 moles (shark) to 8 moles (frog) per mole of protein. Of these BPs shark has the lowest content of tyrosine and the highest contents of lysine, glutamic acid, alanine, and methionine. Since the shark BP appears to be only slightly degraded (see Fig. 6) the composition shown in Table I is probably very close to that of the native protein.

In Table II the amino acid composition of the carp BP preparation is compared with that of a BP of comparable molecular size, the rat S protein. Relative to the rat S protein the carp BP preparation has a much higher content of lysine and a much lower content of histidine. Although it also contains 1 mole of tryptophan per mole of protein, the carp BP preparation contains only one mole of methionine and 0.5 mole of tyrosine per mole of protein. The low content of the latter amino acid suggests that the carp BP is probably degraded. The native carp BP probably contains at least 1 residue of tyrosine and possibly contains an additional methionine residue as well. It is of interest to note that all of the BPs have a high content of the three amino acids, glycine, arginine, and serine, and all contain tryptophan. These characteristics are quite different from those of any of the known histones.The "brain specific histone" isolated from pig brain by Tomasi and Kornguth (32,33), which has a high content of glycine, arginine, and serine, is in all probability myelin basic protein (22).

Chemical studies on BPs have been accompanied by an investigation of their encephalitogenic activity. In collaboration with Dr. E.C. Alvord Jr. of the University of Washington in Seattle and Dr. Seymour Levine of the Bird S.Coler Hospital in New York, we have assayed these proteins in two of the species of laboratory animals most commonly used for studies on EAE-guinea pigs and inbred Lewis rats (Table III).

An interesting difference between the two rat BPs is the marked difference in their ability to induce EAE in the guinea pig.The S protein apparently lacks some structural feature required for this activity. Recent experiments carried out in collaboration with McKneally, Shapira and Kibler (23) have shown that the deletion which differentiates the S from the L protein begins at residue 117 or 118 (just beyond the single tryptophanyl residue at position 116)

TABLE I

Amino Acid Compositions of Myelin Basic Proteins

	Mammalian				Submammalian			
Amino Acid	Human[a]	Bovine[a]	Guinea pig[b]	Rat L[a]	Chicken[c]	Turtle[d]	Frog[b]	Shark[b]
Lysine	6.6	7.6	8.2	7.4	7.0	7.6	6.7	10.2
Histidine	5.6	5.3	5.6	4.5	8.0	6.7	7.0	4.0
Arginine	10.7	10.0	10.1	10.5	11.9	10.7	13.5	9.4
Tryptophan	0.4(1)	0.3(1)	0.3(1)	0.2(1)	0.4(1)	0.1(1)	0.4(1)	0.1(1)
Aspartic acid	7.0	7.4	7.5	7.3	5.6	6.0	7.7	6.1
Threonine	4.6	4.3	5.0	6.0	2.8	2.4	3.8	3.8
Serine	9.2	8.2	8.0	9.9	10.4	9.3	10.3	8.0
Glutamic acid	6.0	7.1	7.3	7.4	5.4	6.6	5.4	10.7
Proline	8.2	7.9	7.6	7.6	7.5	7.7	7.7	5.8
Glycine	15.7	15.2	14.0	14.3	17.2	17.2	12.7	12.7
Alanine	7.9	8.9	8.2	6.9	6.2	7.0	4.8	11.2
Half cystine	0	0	0	0	0	0	0	0
Valine	2.1	1.8	2.0	1.8	3.2	4.1	3.0	3.7
Methionine	1.3(2)	1.2(2)	1.2(2)	1.2(2)	1.2(2)	1.3(2)	1.1(2)	2.7 (5)
Isoleucine	2.1	1.5	2.2	2.0	2.1	2.2	1.3	1.8
Leucine	4.3	5.6	5.0	4.7	4.0	4.2	4.6	4.4
Tyrosine	2.5(4)	2.5(4)	2.4(4)	2.9(5)	1.8(3)	2.8(5)	4.9(8)	1.2(2)
Phenylalanine	5.9	5.1	5.6	5.6	5.5	4.1	5.0	4.4
Arg+Ser+Gly	35.6	33.4	32.1	34.7	39.5	37.2	36.5	30.1

Values are in mole percent. No corrections for hydrolytic losses have been made. Samples were hydrolyzed in constant-boiling HCl containing 4 % thioglycolic acid for 22-24 h at 110° in an evacuated desiccator. Values in parentheses are the estimated number of moles per mole of protein : tryptophan,estimated from ultraviolet absorption spectrum (5) ; methionine,estimated by assuming 170 amino acid residues per molecule ; tyrosine,estimated by assuming 170 amino acid residues per molecule or 2 mole of methionine per mole of protein and by spectral analysis (5). (a) Final purification by gel filtration. Results of a single analysis. (b) Final purification by gel filtration. Average of duplicate analyses. (c) Final purification by ion-exchange chromatography (19). Sample analyzed was most cathodic high pH electrophoretic component (see Fig. 3). Average of duplicate analyses. (d) Final purification by ion-exchange chromatography (19). Sample analyzed was a mixture of high pH electrophoretic components (see Fig.3).Average of duplicate analyses.

TABLE II

Amino Acid Compositions of Rat S and Carp Myelin Basic Proteins

Amino Acid	Rat S[a]	Carp[b]
Lysine	5.4	10.5
Histidine	5.3	2.5
Arginine	13.4	14.1
Tryptophan	0.2 (1)	0.5 (1)
Aspartic acid	8.0	6.0
Threonine	6.6	5.0
Serine	9.6	11.3
Glutamic acid	6.7	7.4
Proline	9.3	8.1
Glycine	12.8	11.7
Alanine	5.2	3.9
Half cystine	0	0
Valine	1.7	4.1
Methionine	1.6 (2)	0.8 (1)
Isoleucine	2.2	2.3
Leucine	4.8	4.6
Tyrosine	1.8 (2)	0.4 (0.5)
Phenylalanine	5.3	6.9
Arg+Ser+Gly	35.8	37.1

Values are in mole percent. No corrections for hydrolytic losses have been made. Samples were hydrolyzed as described in Table I. Values in parentheses are the number of mole per mole of protein estimated as described in Table I. Each protein was assumed to contain 130 amino acid residues per molecule. (a), (b) Same as in Table I.

and ends at residue 157 or 158 (residue numbers correspond to those of the bovine and human BPs). The resulting change in the tryptophan-containing tryptic peptide from Phe-Ser-Trp-Gly-Ala-Glu-Gly-Gln-Lys to Phe-Ser-Trp-Gly-Gly-Arg is probably the reason for the lack of encephalitogenic activity of the S-protein in the guinea pig, since the former peptide isolated from bovine BP and synthesized (8) and a homologous one (Lys replaced by Arg) present in human BP (16), and also synthesized (8), are the only ones reported to contain the major determinants for encephalitogenic activity in the guinea pig. A different peptide (containing tyrosine in place of tryptophan) has been found by Shapira *et al.* (29, 30) and Kibler *et al.* (36) to be encephalitogenic in three other species of laboratory animals : rabbits, squirrel monkeys, and Rhesus monkeys. The particular sequence of amino acid residues in

TABLE III

Susceptibility of Guinea Pigs and Lewis Rats to Mammalian and Submammalian Myelin Basic Proteins

BP	Guinea Pigs	Lewis Rats
Guinea Pig	++	++
Rat L	++	++
Rat S	- (+?)	++
Bovine	++	+
Human	++	+
Monkey	++	+
Rabbit	++	+
Chicken	-	+
Turtle	-	+
Frog	-	+
Shark	-	-
Carp	-	-

BP which is responsible for encephalitogenic activity in Lewis rats has not been defined. Clearly, it cannot be Phe-Ser-Trp-Gly-Ala-Glu-Gly-Gln-Lys, since the rat S protein, which lacks the last half of the sequence, is highly active in this species. Swanborg and Amesse (31) have shown that the tryptophanyl residue is not essential for encephalitogenic activity of bovine BP in the Lewis rat.

The Lewis rat, in some respects, provides an even more interesting tool for bioassay than the guinea pig. The patterns of susceptibility of the two species overlap but are not identical (see Table III). Guinea pigs are susceptible to bovine, human, monkey, rabbit, guinea pig and rat L BPs. They show approximately the same (near maximal) reaction to sensitization by either 5 or 10 μg of each of these proteins. They are much less reactive to rat S BP ; as much as 125 μg produces no clinical and little histologic evidence of disease. Submammalian BPs (chicken,turtle, frog, shark, and carp) were all inactive when either 50 or 250 μg were injected into guinea pigs. Lack of encephalitogenic activity of shark (spiny dogfish) BP in the guinea pig has also been found by H.C. Agrawal _et al_. (37). In contrast, Lewis rats are susceptible to guinea pig and _both_ rat L and S BPs, but are much less susceptible to the BPs of the other mammalian species - bovine, human, rabbit, and monkey. Larger quantities of the latter are required to induce severe EAE in Lewis rats. For example, in a series which involved guinea pig, bovine, human, and rabbit BPs, 2 μg of guinea pig BP induced more severe disease in Lewis rats than 50 μg of each of the other BPs.

In testing submammalian BPs, Dr. Levine found that 50 μg of either chicken or turtle BP induced disease in quantity of bovine BP ; 50 μg of frog BP induced somewhat milder disease. Fifty μg of carp and shark BP induced no EAE in Lewis rats.

Continuation of these studies of the chemistry and biological activity of myelin basic proteins from a wide variety of species, coupled with the use of several test animals, should enable us to trace the evolutionary changes in the molecule and to explain the species-specificity of response to the different encephalitogenic regions.

REFERENCES

1 AGRAWAL, H.C., BANIK, N.L., BONE, A.H., DAVISON, A.N., MITCHELL, R.F. and SPOHN, M., Biochem. J., 120 (1970) 635.
2 CARNEGIE, P.R., Biochem. J., 123 (1971) 57.
3 COTMAN, C.W. and MAHLER, H.R., Arch. Biochem. Biophys., 120 (1967) 384.
4 DEIBLER, G., MARTENSON, R., and KIES, M., 2nd National Meeting of the American Society for Neurochemistry, Hershey, Pa., 1971, Abstr. p.66.
5 EDELHOCH, H., Biochemistry, 6 (1967) 1948.
6 ENG, L.F., CHAO, F.C., GERSTL, B., PRATT, D. and TAVASTSTJERNA, M.G., Biochemistry, 7 (1968) 4455.
7 EYLAR, E.H., Proc. Natl. Acad. Sci. US, 67 (1970) 1425.
8 EYLAR, E.H., CACCAM, J., JACKSON, J.J., WESTALL, F.A. and ROBINSON, A.B., Science, 168 (1970) 1220.
9 FOLCH, J., ASCOLI, I., LEES, M., MEATH, J.A. and LE BARON, F.N., J. Biol. Chem., 191 (1951) 833.
10 GREENFIELD, S., NORTON, W.T. and MORELL, P., 2nd National Meeting of the American Society for Neurochemistry, Hershey, Pa., 1971, Abstr. p.76.
11 KABAT, E.A., WOLF, A. and BEZER, A.E., J. Exptl. Med., 85 (1947) 117.
12 KIES, M.W., Ann. N.Y. Acad. Sci., 122 (1965) 161.
13 KIES, M.W. and ALVORD, E.C. Jr., in Allergic Encephalomyelitis (Ed. KIES, M.W. and ALVORD, E.C. Jr.) Charles C. Thomas, Springfield, 1959), p.293.
14 KIES, M.W., MURPHY, J.B. and ALVORD, E.C. Jr., in Chemical Pathology of the Nervous System (Ed. FOLCH-PI, J.) Pergamon Press, London, 1961, p.197.

15 LAATSCH, R.H., KIES, M.W., GORDON, S. and ALVORD, E.C.Jr., J. Exptl. Med., 115 (1962) 777.
16 LENNON, V.A., WILKS, A.V. and CARNEGIE, P.R., J. Immunol., 105 (1970) 1223.
17 MARTENSON, R.E., DEIBLER, G.E. and KIES, M.W., J. Biol. Chem., 244 (1969) 4261; 4268.
18 MARTENSON, R.E., DEIBLER, G.E. and KIES, M.W., Biochim.Biophys. Acta, 200 (1970) 353.
19 MARTENSON, R.E., DEIBLER, G.E. and KIES, M.W., in Immunological Disorders of the Nervous System , Proc. Ass. Res. Nerv. Ment. Dis., vol. XLIX (Ed. Rowland, L.P.) Williams and Wilkins, Baltimore, 1971, p.76.
20 MARTENSON, R.E., DEIBLER, G.E. and KIES, M.W., J. Neurochem., 18 (1971) 2417.
21 MARTENSON, R.E., DEIBLER, G.E. and KIES, M.W., J. Neurochem., 18 (1971) 2427.
22 MARTENSON, R.E., DEIBLER, G.E. and KIES, M.W., Nature New Biol. (Lond.) 234 (1971) 87.
23 MARTENSON, R.E., DEIBLER, G.E., KIES, M.W., McKNEALLY, S., SHAPIRA, R. and KIBLER, R.F., Biochim. Biophys.Acta,263(1972)193.
24 MARTENSON, R.E. and GAITONDE, M.K., J. Neurochem., 16 (1969) 889.
25 MEHL, E. and HALARIS, A., J. Neurochem., 17 (1970) 659.
26 MORGAN, I.M., J. Exptl. Med., 85 (1947) 131.
27 MORRIS, D., in The Mammals, Harper and Row, New York, 1965, p.184.
28 NAKAO, S. and ROBOZ-EINSTEIN, E., Ann. N.Y. Acad. Sci.US, 122 (1965) 171.
29 SHAPIRA, R., CHOU, F. C-H, McKNEALLY, S., URBAN, E. and KIBLER, R.F., Science, 173 (1971) 736.
30 SHAPIRA, R., McKNEALLY, S.S., CHOU, F., and KIBLER, R.F., J. Biol. Chem., 246 (1971) 4630.
31 SWANBORG, R.H. and AMESSE, L.S., J. Immunol., 107 (1971) 281.
32 TOMASI, L.G. and KORNGUTH, S.E., J. Biol. Chem., 242 (1967)4933.
33 TOMASI, L.G. and KORNGUTH, S.E., J. Biol. Chem., 243 (1968)2507.
34 WAEHNELDT, T.V. and MANDEL, P., FEBS Letters, 9 (1970) 209.
35 WOOD, J.G. and KING, N., Nature, 229 (1971) 56.
36 KIBLER, R.F., RE', P.K., McKNEALLY, S. and SHAPIRA, R., J. Biol. Chem., 247 (1972) 969.
37 AGRAWAL, H.C., BANIK, N.L., BONE, A.H., CUZNER, M.L., DAVISON, A.N. and MITCHELL, R.F., Biochem. J., 124 (1971) 70P.

THE CHEMICAL AND IMMUNOLOGIC PROPERTIES OF THE BASIC A1 PROTEIN OF MYELIN

E.H. EYLAR

The Merck Institute, Rahway, New Jersey 07065 (USA)

Myelin can be considered a simple prototype of a cellular membrane since it appears (42) on electron microscopy to be derived from the plasma membrane of the Schwann cell in the peripheral nervous system and the oligodendroglial cell in the central nervous system (CNS). Based on data (36) from x-ray diffraction and electron microscopy, a repeating pattern was observed in myelin which was compatible with the unit membrane (bimolecular leaflet) concept of membrane structure originally proposed by Davson and Danielli (18).

In order to further our understanding of membrane structure, we chose myelin for isolation of specific protein components that could be solubilized and thereby characterized both chemically and immunologically. This report describes some of our latest results on the A1 protein of CNS myelin, carried out in collaboration with Drs. S. Brostoff, A. Hagopian, J. Caccam, J. Jackson, P. Burnett and G. Hashim. See reference 24 for an earlier review.

Although the myelin membrane is apparently contigous with the plasma membrane, it should be stressed that the composition of these membranes differs markedly (50). When compared to a representative plasma membrane such as that from HeLa cells (7), myelin has a much higher lipid/protein ratio and differs considerably in its lipid composition by its relatively high content of cerebroside and cerebroside sulfate, which are absent in plasma membranes, and α-hydroxy fatty acids which are exclusive to brain. Gangliosides, which are found in plasma membranes, are absent or in trace quantities in myelin. The greatest divergence between myelin and plasma membranes, however, appears in the protein composition. Only three proteins predominate in CNS myelin : the proteolipid protein (50 %),Wolfgram protein (20 %) and the A1 protein (30 %), a basic protein (24) which is the subject of this report. By contrast, plasma membranes contain a large spectrum of proteins (19), some of which may be enzymes; the only enzyme clearly localized in myelin is the 2',3'-cyclic nucleotide 3'-phosphohydrolase (55). Glycoproteins, which are constituents of all plasma membranes (25) are also absent from myelin.

It is readily apparent that in the CNS either the plasma membranes of oligodendroglial cells are highly unusual, or during the conversion process into CNS myelin, pronounced changes occur. The morphogenesis into myelin likely subserves function ; the myelin plays a more static biological role of insulation of the axon in contrast to the active, transporting role of plasma membranes. A major question concerns the biological role of the A1 protein.Since it is not found in other organs, and appears exclusive to myelin, the question arises whether it serves only as a structural element of myelin, or whether it is involved in the assembly of myelin or the interconversion from plasma membrane to myelin membrane. Although direct information is lacking, it can be surmised at the present time that the A1 protein is not present in the oligodendroglial cell membrane since these cells appear resistant to immunologic attack during the course of EAE (51).

Our initial interest in myelin proteins has now extended beyond the application to membrane structure. With the isolation (34,35) of a homogeneous protein which could induce experimental allergic encephalomyelitis (EAE), it was evident that immunologic parameters associated with this disease could be clarified. Although many laboratories had reported on encephalitogenic basic protein (48,66), wide disagreement existed on its composition,properties and biological activity, and it was not until recently that characterizations by conventional methods of protein chemistry were reported (34,35, 58, 62).

Preparation

Detailed procedures for isolation of the A1 protein have been presented elsewhere (34, 60). Prior to biological studies, it was essential to achieve material which was highly purified. A basic protein fraction (34) from bovine spinal cord was eventually prepared containing one major protein, referred to as the A1 protein (80-90 %), and degradative products, referred to as bands A2, A3, and A4 on gel electrophoresis (10-20 %). Proteolysis *in situ* at neutral pH is promoted if the spinal cord is maintained at 4°-37° prior to extraction (34). These observations have been elegantly confirmed by Bergstrand (6), who isolated several peptides produced by the neutral proteolysis *in situ* of the A1 protein, and further demonstrated that reports of several encephalitogenic materials from other laboratories are based on degradation of the A1 protein, the only basic protein in CNS myelin. Alternately, proteolysis may occur during acid extraction,due to an acid protease that is destroyed if chloroform-methanol (2:1) is used to defat the tissue rather than acetone (59). When care is taken to prevent either type of proteolysis, the A1 protein can be prepared in homogeneous form from either whole brain, spinal cord, or myelin. The A1 protein comprises 30 percent of the total myelin protein, and thus is a major myelin component, a figure also found by Eng et al. (21) who used Triton X-100 and high salt for extraction. The A1 protein was prepared in final form by ion exchange chromatography on CM-cellulose. Amberlite IRC-50 or Cellex-P using ionic strength gradients (60).

The A1 protein material appeared homogeneous by gel electrophoresis at pH 4.5 and 8.6, electrophoresis in sodium dodecylsulfate (SDS), ultracentrifugation and immunoelectrophoresis (34). At pH 10.5, gel electrophoresis shows a minor degree of microheterogeneity with approximately 10-20 percent of the material migrating as two slower bands, an observation essentially in agreement with that of Martenson et al. (57). Although considerable emphasis has been given this phenomenon (56), we consider the microheterogeneity trivial. It is a usual feature of most proteins, possibly arising from deamidation occurring during acid extraction.

Chemical Properties

The physico-chemical properties of the A1 protein are shown in Table I. The molecular weight, as determined by the amino acid sequence, is near 18,500 for both the human and bovine A1 proteins. The variation in molecular weight between ten mammalian species appears small because these proteins migrate identically on gel electrophoresis in SDS (unpublished observation). The molecular weight determined by other techniques, including tryptophan determination, agrees with the sequence value.

The most important feature of the A1 protein is the open conformation, first shown by viscosity studies (35) which revealed an intrinsic viscosity of 9.3 dl/g, a value since confirmed (15).Based on the viscosity data, an axial ratio of 10:1 was calculated for the A1 protein. Unlike most proteins which are globular in aqueous media, the A1 molecule appears unfolded. Moreover, the data in Table I reveal that the A1 protein lacks significant secondary or tertiary structure. Even treatment with 8 M urea or heating at 100° does not limit the ability of the A1 protein to induce EAE or combine with antibody (35). Thus, the A1 protein is stable to denaturation because it is "denatured" to begin with. At this stage of our investigation, in 1968, it was evident that tertiary or secondary structure might be less important than the primary structure

TABLE I

Physico-Chemical Properties of the A1 Protein

Property	Method
Molecular weight :	
18,395 (Bovine); 18,625 (Human)	Sequence
18,600- 20,100 (Bovine); 19,400-19,600 (Human)	Ultracentrifugation
19,500 (Bovine); 19,500 (Human)	Gel electrophoresis
19,500 (Bovine);	Gel filtration
18,200 (Bovine); 17,400 (Human)	Tryptophan analysis
Secondary structure :	
α-helix None	Circular dichroism;
β-structure None	Optical rotatory dispersion
Open conformation :	
Intrinsic viscosity, 9.3 ml/g	Viscosity
Axial ratio, 10:1	Viscosity
Resistant to denaturation	100° for 1 h ; 8 M urea
Highly vulnerable to proteolysis	Trypsin, pepsin, chymotryptic digestion
High proline and glycine	Amino acid analysis

in defining the biological activity of the A1 protein, and the need for the amino acid sequence was thus emphasized.

Amino Acid Sequence

The sequence of both the human (13, 23) and bovine A1 proteins (27) is shown in Fig. 1. The amino acid sequence was assembled primarily from tryptic and peptic peptides (27) ; CNBr and BNPS-skatole reagents were used to cleave the COOH-methionyl and COOH-tryptophanyl linkages respectively giving peptides (10, 32, 43) that were very helpful in establishing the sequence. Twenty-seven tryptic peptides were isolated from the bovine A1 protein following chromatography on Dowex 50 or peptide mapping ; 16 peptic peptides were purified by Cellex P chromatography. Most of the tryptic peptides could be accounted for by peptide mapping as shown in Fig.2. The sequences of all peptides were determined by the direct and subtractive Edman degradation procedures (27).

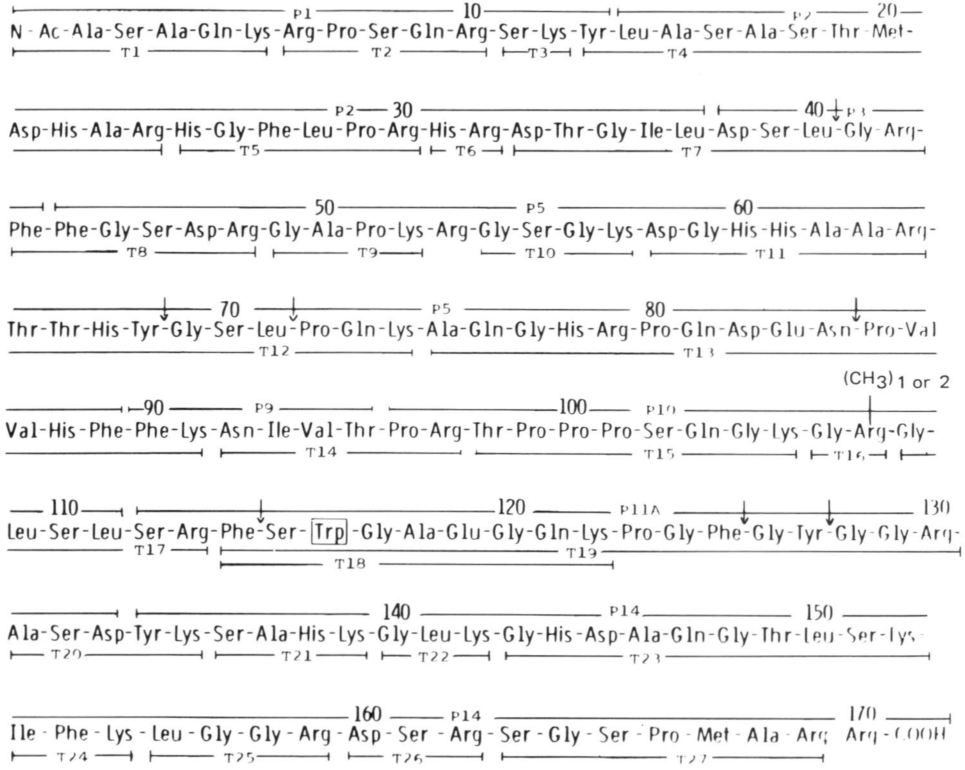

Fig. 1. The amino acid sequence of the bovine A1 protein is shown with the tryptic peptides designated by T and the peptic peptides by P. The sequence of the human A1 protein is identical except for substitutions as shown above the bovine sequence. Additional bonds hydrolyzed by pepsin are denoted by arrows. The single tryptophan residue 116 is enclosed in a box. Arginine residue 107 exists primarily as the monomethylarginine and dimethylarginine derivatives.

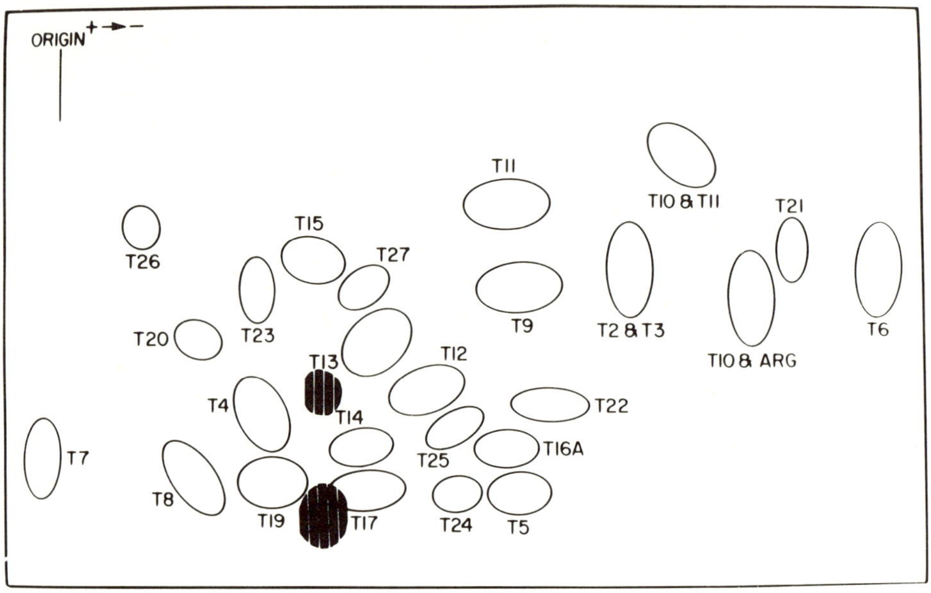

Fig. 2. The peptide map of tryptic peptides derived from the bovine A1 protein is shown.The two radioactive forms of peptide T15, containing C^{14}-N-acetylgalactosamine in glycosidic linkage with threonine residue 98, are shown by cross hatches.

The A1 protein was found to have a unique sequence that showed no obvious similarity to other basic proteins such as histone IV, lysozyme, cytochrome c or ribonuclease. No periodicity of basic residues was found, but rather a general distribution existed over the polypeptide chain. Several peptide segments of nine residues exist, however, in which basic residues are absent ; the somewhat nonpolar character of these regions suggests a possibility for participation in nonpolar interactions. The sequence suggests, in fact, that the A1 molecule is ideally designed for maximal interaction, both electrostatic and nonpolar, with phospholipids or other components of the myelin membrane. Although nonpolar interactions between proteins and lipids in membranes have been generally emphasized (37), it is very likely that the A1 protein binds electrostatically to phospholipids, particularly in view of the open conformation of the A1 molecule as emphasized by the 25 net positive charges, the proline residues and the relatively minor content of nonpolar residues. Within the myelin framework, however, it is possible that the A1 molecule assumes some folded conformation upon interaction with either lipid components or other proteins such as proteolipid. Numerous studies on myelin support a structure compatible with the unit membrane model (36). The properties of the A1 molecule are actually similar to the extended protein originally postulated by Danielli (18) to be located at the charged surface of the phospholipid leaflet. It would be naive, however, at this stage to presume such a simple orientation for the A1 molecule in view of the possible influence of the lipids and other proteins present in myelin.

Structural Features

Near the middle of the A1 molecule several unusual structural characteristics were found; (42) the methylated arginine residue 107; (36) the proline-rich region Pro-Arg-Thr-Pro-Pro-Pro ; (18) the site of glycosylation, threonine residue 98. These features are remarkable because they are not commonly found in proteins, and thus may be related to the unique biological role of the A1 protein in the myelin membrane.

Methylated arginine. - Arginine residue 107 of the A1 protein exists primarily in two modifications, as dimethyl- and monomethylarginine (8). Direct isolation of these residues from the bovine A1 protein was obtained, following pronase and amino peptidase M digestion, by chromatography on Dowex 50 W. Final purification (Fig.3) of the N^G-monomethyl arginine was achieved by preparative paper chromatography in pyridine-acetone-3M NH_4OH (10:6:5). N^G-dimethylarginine was purified by high voltage elctrophoresis at pH 2.7 (Fig.3). The dimethyl derivative is easily detected on the Beckman amino acid analyzer (Fig. 4) where it elutes prior to arginine as shown in Fig. 2 ; the elution of the monomethylarginine (44 min) is obscured, however, because it nearly coincides with arginine (45 min). The structures of the isolated derivatives were deduced from the alkaline degradation products ; the monomethyl derivatives giving ornithine, citrulline and methylamine, and the dimethyl derivative giving dimethylamine in addition (8). The finding of both methylamine and dimethylamine in the alkaline hydrolyzate of the dimethylarginine fraction reveals the presence of two forms : N^G,N^G and N^G,N'^G-dimethylarginine. Combined gas chromatography- mass spectroscopy determinations gave a ratio of 1:2 respectively for these derivatives (9).

Localization of the methylated arginine residues in the A1 molecule were accomplished by the examination of the tryptic and peptic peptides shown in Table II. When the bovine A1 protein is digested with trypsin and peptides mapped (Fig. 2), two peptides (peptides T 16A and T 17) were found which are derived from the same region of the A1 molecule. It became apparent that the Arg-Gly

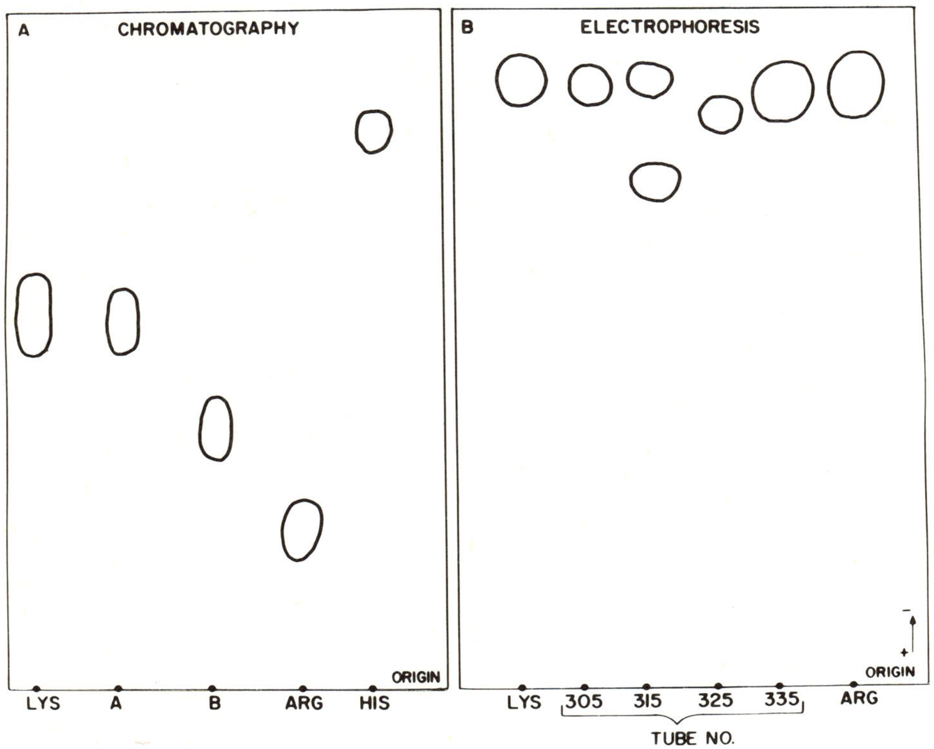

Fig. 3. Chromatography and electrophoresis.(A) ascending chromatography in the solvent pyridine-acetone-3 M NH_4OH (10:6:5). Fraction A is identical with dimethylarginine ; fraction B is identical with monomethylarginine. On paper electrophoresis (B), dimethylarginine, in tube no. 315, migrates more slowly than monomethylarginine (tube no. 325) arginine, or lysine.

TABLE II

Peptides Isolated from the Methylated Arginine Region of the A1 Protein

Peptide	Sequence
	(Methyl) 1 1 or 2
Tryptic Peptide T 16 A (Res. 106-113)	Gly-Arg-Gly-Leu-Ser-Leu-Ser-Arg
Tryptic Peptide T 17 (Res. 108-113)	Gly-Leu-Ser-Leu-Ser-Arg
Peptic Peptide P 10 (Res. 96-111-	Pro-Arg-Thr-Pro-Pro-Pro-Ser-Gln-Gly-Lys-Gly-Arg-Gly-Leu-Ser-Leu
Peptic Peptide P 10A (Res. 96-111)	Pro-Arg-Thr-Pro-Pro-Pro-Ser-Gln-Gly-Lys-Gly-Arg-Gly-Leu-Ser-Leu 1
	$(\text{Methyl})_2$ 107
Summary :	----Gly-Arg-Gly---- | Site of methylation

linkage, found in peptide T 16A, is normally cleaved giving peptide T 17. When residue 107 is occupied by methylated arginine derivatives, however, the action of trypsin is partially inhibited. All normal Arg-Gly sequences are hydrolyzed under the conditions used. The dimethylarginine at position 107 accounts for 20-60 percent of the arginine variants in peptide T 16A depending on species (Table III) , the rest being the monomethyl derivative. With the bovine protein, peptides T 16 A and T 17 are found in a 3:2 ratio. Thus the dimethyl derivative accounts for 0.2 mole/mole A1 protein at position 107, a value close to that found by direct hydrolysis of the A1 protein. The monomethyl derivative must occupy, therefore, at least 40 percent of the A1 molecules at this position and possibly as much as 80 percent,depending on its trypsin resistance. Some nonmethylated arginine probably exists there as well because the A1 protein serves as an acceptor for the methylase from brain using S-adenosylmethionine (4). The figure reported by Baldwin and Carnegie (4) of 3:5 for the ratio of monomethyl to dimethyl-arginine is clearly incorrect because it is based only on the ratio found in tryptic peptide T 16 A where the dimethyl derivative would be expected to predominate because of its greater resistance to tryptic hydrolysis.

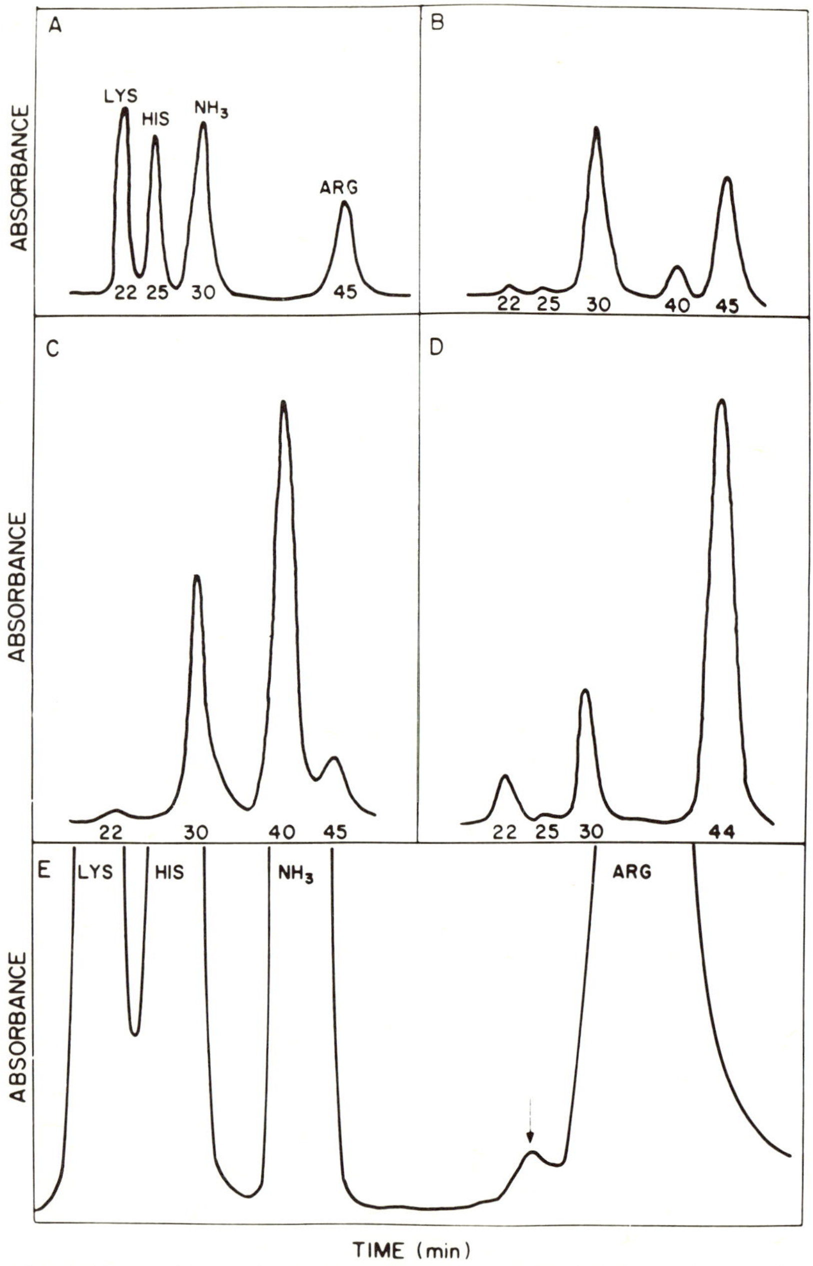

Fig. 4. Amino acid analyses performed on the basic column of a Beckman autoanalyzer. (A) standard run of basic amino acids and ammonia; (B) hydrolysis of peptide T 16A (see Fig. 1) ; (C) dimethylarginine plus ammonia ; (D) monomethylarginine plus ammonia ; (E) expanded run of a hydrolyzate of the bovine A1 protein (overloaded to show dimethylarginine at the arrow).

TABLE III

Amino Acid Composition of Tryptic Peptides T 16A Containing Dimethylarginine from Various Species[†]

	Bovine	Chicken	Guinea pig	Human	Monkey	Rabbit	Rat
Lys	-	1.0[††]	-	-	-	-	-
$Arg_{(Methyl)_2}$	0.3	0.4	0.2	0.6	0.4	0.3	0.3
Arg	1.4	1.0	1.3	1.4	1.6	1.5	1.8
Thr	-	1.0	-	-	-	1.0	-
Ser	2.0	1.2	2.1	1.9	1.9	1.8	1.8
Gly	1.6	1.7	1.8	1.8	2.0	2.3	1.8
Leu	2.0	2.0	2.0	2.0	2.0	2.0	2.0

[†]The turtle A1 protein also contained dimethylarginine at the same position, 0.2 mol/mol of peptide, as determined from a chymotryptic peptide analogous to peptide P 10A. [††]Ratios normalized to Leu (2.0).

The importance of the methylated arginine residue is emphasized by the phylogenetic distribution shown in tryptic peptide T 16A (Table III). All species examined including chicken and turtle show dimethylarginine at a position analogous to that in the bovine A1 protein.

The methylation of arginine residue 107 provides a hydrophobic contribution that could stabilize a crucial interaction either within the A1 molecule or with other myelin components. It appears highly specific and therefore relevant to the function of the A1 protein for many reasons : it is rare in proteins, occuring possibly in histones in addition to the A1 protein (61) ; it occurs at only one of the 19 arginine residues in the A1 molecule and it occurs in the A1 proteins from all species examined as well as in basic protein from peripheral nerve (unpublished observation).

<u>The proline-rich region</u>. - Located near the methylated arginine residue is another rare region, a proline-rich region which includes a triproline sequence :

```
       96              101
-Pro-Arg-Thr-Pro-Pro-Pro-Ser
```

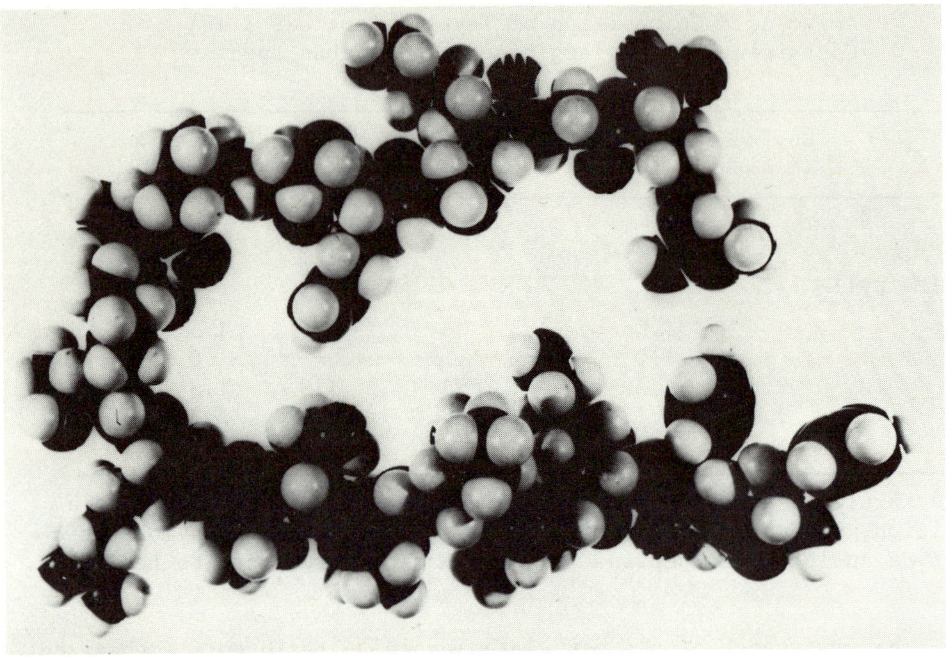

Fig. 5. A possible conformation for the region of the A1 protein containing the methylarginine locus at position 107 (large arrow): Phe-Phe-Lys-Asn-Ile-Val-Thr-Pro-Arg-Thr-Pro-Pro-Pro-Ser-Gln-Gly-Lys-Gly-Arg (Methyl). The three small arrows point to the Pro-Pro-Pro sequence. The two phenylalanine side chains occur directly across from the methylarginine residue in this conformation.

A possible conformation for this region is shown in Fig. 5. It is evident that severe conformational restrictions are imposed by the proline residues that could produce the sharp U-shaped bend as shown ; the A1 protein would thus assume an open double chain structure. Viscosity studies reveal a 10:1 axial ratio, a figure more compatible with a double chain structure than a random coil. The proximity of the proline bend region to the methylated arginine at residue 107 suggests an action in concert to stabilize the A1 polypeptide chain. We speculate that stabilization might occur by cross-chain interaction of the methylated arginine with the proximal phenylalanine side chains.

<u>The site of glycosylation</u>. - Of a large number of proteins tested (38) only the A1 protein served as a natural acceptor for the polypeptide : N-acetylgalactosaminyl transferase from sub-

maxillary glands (39), an enzyme responsible for formation of a sugar-protein linkage as the crucial first step in the biosynthesis of certain glycoproteins. The reaction involves the formation of a glycosidic linkage as shown :

$$\text{UDP-GalNAc} + \text{A1 protein} \rightarrow \text{A1 protein-O-GalNAc} + \text{UDP}$$

It was found that threonine served as the specific attachment site for the N-acetylgalactosamine residue. Once the sequence of the A1 protein was determined, the problem was reinvestigated (41). The A1 protein was incubated with the transferase and a relatively large quantity of UDP-GalNAc-^{14}C in order to form sufficient amounts of A1 protein-GalNAc-^{14}C product. Following trypsin digestion, most of the radioactivity could be localized in only two radioactive peptides shown on the peptide map (Fig.2). These peptides were isolated ; as shown by their sequences (Table IV), it is evident that

TABLE IV

Identification of the Acceptor Region in the A1 Molecule Recognized by the Polypeptide : N-Acetylgalactosaminyl Transferase

Radioactive peptides :

```
                                          GalNAc-C14
                                              |                          105
(Peptide T 3-C14)Asn-Ile-Val-Thr-Pro-Arg-Thr-Pro-Pro-Pro-Ser-Gln-Gly-Lys

(Peptide T2-C14)                GalNAc-C14-Thr-Pro-Pro-Pro-Ser-Gln-Gly-Lys

Rabbit IgG region : -Lys-Pro-Thr-Cys-Pro-Pro-Pro-
                                 |
                               GalNAc
```

(In the original, Thr-Pro-Pro-Pro in both peptides and Cys-Pro-Pro-Pro- in the IgG region are underlined.)

Evidence for the threonine (no.98) as the acceptor residue of peptide T2-C^{14} :

1. Only threonine significantly destroyed (32 %) by alkaline borohydride treatment.

2. After one step of the Edman degradation, over 80 % of the Gal NAc-C^{14} was removed.

they are both derived from the same region of the polypeptide chain, the proline-rich region. The amino acid-sugar linkage was shown unequivocally to occur at threonine residue 98 giving Thr-O-GalNAc-^{14}C (Table IV).

These data establish that only one residue is glycosylated in the A1 protein. It was concluded that the peptide region surrounding the threonyl 98 locus must constitute a highly specific site for recognition by the transferase, a site which likely contains the essential residues for acceptor recognition during the biosynthesis of mucins and other glycoproteins. A stiking comparison exists with the hinge region of rabbit IgG (69) where a GalNAc-O-Thr linkage is found proximate to a proline-rich region. We propose, therefore, that a Thr-X-Pro or Thr-X-Pro-Pro sequence may be essential for acceptor activity. However, this is probably not the only requirement. The 22 residue peptic peptide containing this region, peptide P9-P10, also has receptor activity ; 40 % as high as the A1 protein (40). From these data, it is highly probable that like the immunologic properties of the A1 protein, the acceptor activity is dependent on localized tertiary structure over a short segment of the polypeptide chain and not necessarily on some special tertiary arrangement of distant segments of the A1 polypeptide chain.

The question arises whether the glycosylation produced *in vitro* with the A1 protein is fortuitous since it is not a glycoprotein as isolated. It is possible that sugar residues are attached at some stage prior to its assimilation into myelin and ultimately removed (22).

The Induction of EAE

Many components of CNS tissue had been proposed as encephalitogenic agents including proteolipid, collagen-like protein, glycolipopeptides and phospholipids. The low level of encephalitogenic activity found in these materials could be accounted for by contamination with A1 protein (54). Encephalitogenic basic peptide material, molecular weight below 5000, derived by dialysis of defatted CNS tissue, was also reported (54,65). Since the A1 protein appeared to be the only basic protein in myelin (34) in contrast to reports of 5-6 basic components (2), and since degradation occurs readily (34), it was logical to assume that the active peptides were actually derived by proteolysis from the A1 protein. This interpretation was in keeping with the open conformation of the A1 protein suggesting that the antigen(s) in the A1 protein responsible for EAE resides in a small region of the polypeptide chain. Since our main objective has been to identify the encephalitogenic antigens, our strategy has focused on peptides derived from the A1 protein.

Encephalitogenic peptides. - We reported (31) in 1968 the sequences of two peptic peptides which were highly encephalitogenic in guinea pigs. These peptides (peptides E and E1) were derived from the same region of the A1 protein ; the region surrounding the single tryptophan residue (Tables V and VI). The peptides appeared nearly as encephalitogenic on a molar basis as the original A1 protin. It was evident that a major encephalitogenic determinant existed within the 14 residues of the peptide E. In view of the specificity of immunologic reactions, the peptide presumably maintains a close spatial similarity to the same site in the intact A1 protein, thus providing strong support for the open conformation of the A1 protein. The high encephalitogenic activity of the tryptophan peptide was later confirmed by Carnegie who isolated the identical peptide from a peptic digest of human A1 protein (11).

TABLE V

Encephalitogenic Activity of Synthetic Peptides

Peptide	Encephalitogenic	Sequence
T 18	+	Phe-Ser-Trp-Gly-Ala-Glu-Gly-Gln-Lys
E	+	Ser-Arg-Phe-Ser-Trp-Gly-Ala-Glu-Gly-Gln-Lys-Pro-Gly-Phe
S 1	+	Ser-Arg-Phe-Ser-Trp-Gly-Ala-Glu-Gly-Gln-Lys
S 2	-	Ser-Arg-Phe-Ser-Trp-Gly-Ala-Glu-Gly-Gln
S 3	+	Ser-Arg-Phe-Ser-Trp-Gly-Ala-Glu-Gly-Gln-Arg
S 4	-	Ser-Arg-Phe-Ser-Trp-Gly-Ala-Glu-Gly-Gln-Ile
S 5	-	Ser-Arg-Phe-Ser-Trp-Gly-Ala-Glu-Gly-Ile-Lys
S 6	+	Ser-Arg-Phe-Ser-Trp-Gly-Ala-Ile-Gly-Gln-Lys
S 7	-	Ser-Arg-Phe-Ser-Phe-Gly-Ala-Glu-Gly-Gln-Lys
S 8	-	Ser-Arg-Phe-Ser-Val Gly-Ala-Glu-Gly-Gln-Lys
S 9	+	Ser-Arg-Phe-Ala-Trp-Gly-Ala-Glu-Gly-Gln-Lys
S 10	+	Ser-Arg-Val-Ser-Trp-Gly-Ala-Glu-Gly-Gln-Lys
S 11	±	Phe-Ser-Trp-Gly-Ala-Gly-Gly-Glu-Lys
Required sequence		--------Trp----------------Gln-Lys

TABLE VI
Immunological Properties of Peptides and Modified Derivatives of the A1 Protein

Peptide or derivatives	Residues	EAE induction in†	PHI test††	Skin test†††	Capillary migration test††††
E	112-125 (14)	Guinea pig + Rabbit +	Neg.		65
E1	112-133 (22)	Guinea pig + Rabbit +	Neg.		
T18	114-122 (9)	Guinea pig + Rabbit + Monkey -	Neg.		60
R	44-89 (46)	Guinea pig + Rabbit + Monkey ?	Neg.		
CBI	1-22 (20)	Guinea pig ±	Neg.		
T	117-170 (54)	Guinea pig - Rabbit - Monkey +	Neg.	+	
L	1-117(117)	Guinea pig - Rabbit +	+	+	
HNB-A1 protein	1-170(170)	Guinea pig - Rabbit + Monkey +	+	+	35
Oxidized A1 protein	1-170(170)	Guinea pig + Rabbit +	+	+	
A1 protein (Bovine)	1-170(170)	Guinea pig + Rabbit + Monkey +	+	+	40

†Experimental allergic encephalomyelitis (EAE) evaluated by clinical and histologic evaluation (33,34).The (-) sign refers to activity which is 10-100 fold less than the A1 protein on a molar basis. ††Measures the ability to inhibit antibody to the A1 protein in the passive hemagglutination of erythrocytes (containing adsorbed A1 protein). †††Measured as the area of erythema in guinea pigs (sensitized 9 days earlier with A1 protein) produced after 24 h when injected with 5-50 μg of test material intradermally. ††††Data represents percent migration of macrophages in chambers with antigen compared with percent migration in chambers without antigen. The peritoneal exudates used in this study were taken from guinea pigs 7 days after sensitization with either peptide or A1 protein derivative. Percent migration in controls tested with albumin or γ-globulin was 85-95. Some of these data are from Spitler et al. (68).

It has been reported that tryptic digestion of the basic protein greatly reduced encephalitogenic activity in guinea pigs (14, 44, 49). We isolated, however, a tryptic peptide (peptide T 18), having only nine residues, which was highly active (30). Significantly, tryptic peptide T 18 was derived from the same tryptophan region as the active peptic peptides (Table V). Interestingly, encephalitogenic peptide T 18 is far less effective when present in the whole tryptic digest than when given alone (30, 44).

Tryptic peptide T 18 established the length limit of the active region as nine residues. In order to specify the essential residues within the framework of peptide T 18, a series of peptides were synthesized by the Merrifield solid state technique (70). The encephalitogenic activity of synthetic peptide S1 equalled the activity of the derived peptic or tryptic peptides, and virtually confirmed the determined sequence. The results (Table V) showed that at least three residues are required for EAE induction, the tryptophan, glutamine and lysine. The tryptophan was not replaceable by phenylalanine, another aromatic residue. Lysine was replaceable with arginine, another basic residue, which occurs at position 122 in the human sequence. Deamidation of the glutamine residue (peptide S 11) leads to inactivation as judged clinically but still produced histological lesions at 3 μg doses.

In summary, it appears that only one major determinant exists in the A1 protein which is responsible for EAE in the guinea pig. Two other regions (12, 46) mentioned as possible antigenic sites, residues 1-22 and 44-89, have only negligible activity in our laboratory (26). The smallest active peptide segment is nine residues, identical with peptide T 18. If the NH_2-terminal Phe is removed, the activity is greatly reduced presumably because of the proximity of the terminal amino group to the tryptophan residue. Of the three essential elements, the tryptophan residue is of most interest because it occurs only once in the entire A1 sequence. Thus the use of synthetic peptides permitted definition of the essential elements of the tryptophan region in disease induction in guinea pigs.

Modification of the tryptophan residue. - The agent 2-hydroxy-5-nitrobenzyl (HNB) bromide is ideal for modifying the tryptophan residue of proteins because the reaction is specific and complete (5). When the single tryptophan residue of the A1 protein is modified, the protein becomes virtually inactive in guinea pigs, as first reported in 1969 (32), when tested at 150 μg (Table VI) ; no clinical signs were found, and only a paucity of histologic signs (29, 44). These results have now been confirmed elsewhere (16) and offer strong evidence that the tryptophan site is the major encephalitogenic antigen in the A1 protein when tested in guinea pigs.

Cleavage at the tryptophan residue. - Further elucidation of the structure responsible for EAE induction was made possible by using peptides derived from the A1 protein by chemical cleavage of the COOH-tryptophanyl linkage with N-bromosuccinimide (32) or BNPS-skatole (10). The latter agent is preferable because it is more specific ; cleavage at other residues such as tyrosine and histidine does not occur. Approximately 40 percent of the A1 molecules were cleaved yielding two peptides, referred to as peptides L (1-117) and peptide T (118-170), which were subsequently isolated. As expected from cleavage of the tryptophan region, these two peptides were relatively non-encephalitogenic (Table VI). It should be noted, however, that the oxidized A1 protein, prepared by oxidation of position 2 of the indol ring of tryptophan (with BNPS-skatole) under mild conditions where cleavage is negligible, was still fully active (10) (Table VI). Thus, unlike the case of the HNB modification, where the bulky group adds to the tryptophan across positions 2 and 3 of the indol ring, oxidation at position 2 does not violate the critical participation of the tryptophan residue in eliciting the immunologic response leading to EAE.

Peptides T and L provide useful means for segregation of immunologic properties of the A1 protein. In the passive hemagglutination inhibition (PHI) test, antibody, if premixed with A1 protein, is prevented from agglutinating chicken erythrocytes to which A1 protein is adsorbed. It was found that peptide L, but not peptide T,was very active in the PHI test (10), approximately equal in activity to the A1 protein (Table VI). It was concluded that a major antigenic site, reactive with humoral antibody, resides in the peptide L region. In contrast, peptides from the tryptophan region were inactive in the PHI test. Similar results in the PHI test were obtained with antisera from sheep, rabbits, guinea pigs and monkeys (42).

These results may help to delineate between cell-mediated and humoral factors in EAE with regard to both disease induction and suppression. It is still uncertain whether the immunologic events associated with EAE are entirely due to cell-mediated phenomena or whether circulating factors are also involved (1). It is clear that within the A1 molecule the encephalitogenic determinant does not coincide with the antigenic region associated with antibody as measured by the PHI test. Moreover, this antibody may be associated with blocking or suppression of the encephalitogenic response (67). These data are consistent with results obtained with the HNB-A1 protein, which was fully active in the PHI test, and shows that the antibodies responsible for the PHI test do not require the tryptophan region, but are primarily directed toward some other site of the A1 molecule (29, 42).

Delayed hypersensitivity. - Cell mediated immunologic events have generally been emphasized in EAE because of the close association with delayed hypersensitivity and disease development (1). We showed (34) a parallel in development of histologic lesions in EAE (perivascular cuffing) and the delayed-type skin response;both appeared five days after sensitization and both increased in severity until the twelfth day. The correlation between the delayed-skin test and disease development is not a necessary prerequisite for EAE, as previously suggested (1), since the two phenomena can be dissociated as shown in Table VI. The HNB-A1 protein, which does not induce EAE, nevertheless gives a delayed hypersensitive response, as strong as the A1 protein as measured by skin-test, lymphocyte stimulation and macrophage migration-inhibition (68). These data reveal that other regions of the A1 molecule, unlike the tryptophan region, are capable of inducing delayed hypersensitivity but not disease in guinea pigs. This result was confirmed using peptides L and T, both of which give a definite delayed-skin reaction, but which do not induce EAE. If EAE is primarily due to cell mediated phenomena, it is not clear why some antigenic sites, which induce sensitized lymphocytes, do not lead to the disease state. Discrimination at the target tissue, or alternatively, the relative number of sensitized cells to a particular antigen might be factors to explain these results ; but humoral factors might also be involved.

As expected, the tryptophan region as defined by synthetic peptide S1 induces a delayed hypersensitive reaction in guinea pigs (Table VI) as measured by the MIF test, which is produced in response to either the peptide or the A1 protein. A positive delayed-type skin test is also induced by the peptide, but it is necessary to employ large concentrations of peptide, presumably because of its more rapid diffusion from the skin site.

Other antigenic sites responsible for EAE. - Once it was established that the tryptophan region was primarily responsible for EAE in the guinea pigs, two questions arose : does this region account exclusively for the encephalitogenic activity of the A1 protein in the guinea pig, and is it active in other species? Although other regions of the A1 molecule may have trace activity in the guinea pig, it is clear that the tryptophan region, as defined by the nine residues of peptide T 18 is the major site since : (a) it is nearly as active on a molar basis as the intact A1 protein ;

(b) the HNB-modification of the A1 protein is several hundred fold less active ; and (c) A1 proteins of all mammalian species tested thus far including bovine, rat, mouse, dog, human, monkey, horse, guinea pig and rabbit have the same activity in guinea pigs (42) and the same sequence in the tryptophan region (except for Lys to Arg changes in human and rat A1 proteins). Only the chicken and turtle A1 proteins, where the sequence is modified, are nonencephalitogenic. Thus the phylogenetic variation as tested in guinea pigs is consistent with the structural requirements established with the synthetic peptides.

It has been proposed (12) that the peptide with residues 1-20 derived by CNBr treatment of the A1 protein is encephalitogenic in guinea pigs. In our laboratory, we have not been able to confirm this result. In any case, a recent report (52) now shows that it has insignificant activity, 500 fold less active than the tryptophan region.

We recently (26) derived a peptide, referred to as peptide R, from the A1 protein in high yield that is active in rabbits, but not active in guinea pigs. This peptide appears to be identical with that derived from spinal cord degraded during acid extraction, which was shown to be active in rabbits (46). The acid protease (59) responsible for the degradation, is similar to pepsin and appears to show a preference for the Phe-Phe linkages which are cleaved at each end of the peptide R molecule. Thus peptide R, which is not significantly active in guinea pigs, contains a peptide region which is encephalitogenic in rabbits. In the A1 molecule, therefore, it appears that two major antigens exist which induce disease in rabbits. It should be noted that peptide R does not combine in the PHI test with rabbit antibody to the A1 protein, again showing that encephalitogenic sites do not necessarily represent an antigenic focus for humoral antibody (Table VI).

It was reported by Kibler *et al*. (47) that peptide R contained 10 percent contamination. A critical factor in the study of encephalitogenic peptides is purity. Because of the limitations in the EAE assay, and the high activity of some peptides, it is essential that strict precautions be taken to insure the absence of contaminants. Even 5 percent contamination could prove misleading.

EAE in monkeys. - Kabat *et al*. (45) established over 20 years ago that the monkey is an ideal animal for study of EAE induced by whole CNS tissue. Because of the relevance to human demyelinating disease, we investigated EAE induction in the monkey using the purified human A1 protein ; a consistent course of clinical and histological signs of EAE was found not unlike that seen when whole CNS tissue is used (33). Clinical signs develop on days 13-25 ; the histologic lesions are characterized by infiltrations of polymorphonuclear cells as well as lymphocytes,in contrast to guinea pigs where the polymorphonuclear cells were rarely observed. It is likely that the ratio of A1 protein to tubercle bacillus in the Frend's adjuvant may influence the timing and severity of EAE development ; the ratio was 1,000 fold larger in the monkeys than in the guinea pigs.

In order to ascertain the antigenic regions of the A1 protein responsible for EAE in the monkey (28), a series of peptides were tested for encephalitogenic activity (Table VI). Neither the tryptophan peptide nor peptide R was appreciably active, but peptide T was very active, equal to the A1 protein on a molar basis. Peptide R showed some activity at high concentration, but interpretation must be made with caution, because of the possibility of trace contamination with a highly active peptide. The HNB-A1 protein is encephalitogenic in monkeys, thus confirming the inactivity of the tryptophan region (Table VI).

The activity of peptide T suggested that, in monkeys, the main encephalitogenic site resided in the COOH-terminal region. This result was confirmed with the finding that peptide P14, a 37-residue peptide beginning with tyrosine residue 134, was also very active (Fig. 1). By analogy to the tryptophan region, it can be anticipated that a small region of perhaps 6-10 residues will be found in this region as the major diseaese-inducing site. There is no obvious similarity of the sequence in peptide P14 to either the tryptophan or peptide R regions. These data further emphasize the phylogenetic variation in the immunologic response leading to EAE. Because of this variation, there is no assurance that the disease-inducing site in humans is the same as in monkeys. It is apparent that this antigen is present in the human, monkey, bovine and rabbit A1 proteins since all induce EAE in the monkey. Based on the results of postrabies encephalitis in man, which can be considered very similar to EAE, it is apparent that the specificity for disease induction in humans is present in both the rabbit and mouse A1 proteins.

Blocking and Suppression of EAE with A1 Protein

Several studies (3, 17, 20) in guinea pigs have shown that if basic protein material (not always characterized was given prior or near the times of sensitization, then EAE induction was inhibited. The mechanism for blocking EAE is not understood ; it could involve either induction of toleráce or the production of blocking antibody (67). We have also studied the blocking of EAE in guinea pigs (antigen given prior to sensitization) as well as suppression (antigen given after sensitization). It is surprising that the HNB derivative, which is not encephalitogenic, is nearly as effective as the protein in suppressing EAE when given 5-7 days after sensitization.

In order to achieve conditions more clinically relevant, we used human A1 protein to suppress EAE in monkeys at the most critical stage in disease development, after clinical signs had appeared. As Table VII shows, thirteen of sixteen animals survived when the A1 protein or HNB derivatives was administered within 24-48 hours after unmistakable clinical signs had been observed ; these signs included ataxia, hind leg weakness, uncoordination and lethargy. Loss of appetite always occurred 2-5 days earlier. All 20 control animals died within 1-3 days ; 2 of 3 animals of the suppressed group survived 10 and 23 days before dying. It should be noted that in the 13 animals which survived, the clinal signs disappeared, and the animals appeared normal.

TABLE VII

The Suppression of EAE in Rhesus Monkeys Following Appearance of Clinical Signs

Suppressive† treatment	Total No. of animals	No.developing clinical signs	No.dead	No.recovered
A1 protein	12	12	3	9
HNB-A1 protein	4	4	0	4
Histone	10	10	10	0
None	10	10	10	0

†EAE was induced in all monkeys by injection into the footpad of a total of 5 mg A1 protein (human) in complete Freund's adjuvant. Suppressive treatment, with the material indicated, was begun when unmistakable clinical signs of ataxia, weakness or paralysis, and loss of coordination had appeared. Anorexia generally preceded these signs by 2-4 days. Control animals died in 1-4 days after general clinical signs were first observed.

In all cases, the animals were given procaine penicillin G (75,000 units/day), and when not eating were force fed twice daily with A-B-dextrose solution (40 ml total), Allstate Veterinary Supply Company.

Two major points emerge from this study : a short-term suppression in which improvement in the clinical state was observed within 24 h after the first doses,and a long-term effect in which permanent suppression was observed as long as suppressive injections (2 mg/day) are sustained daily beyond a 10-12 day period. If injections were stopped at any time during this critical period,then full expression of clinical signs of EAE reappeared after a 3-4 day lapse. Even after 7-9 days of injections, at which point the monkeys had returned to an apparently normal state, cessation of injections led eventually to a return of severe clinical signs.However, in these animals also, the clinical course of the disease was again reversed, and a normal state was attained when suppressive treatment was resumed.

The most probable mechanism of suppression appears to be a direct destructive action of the injected antigen on the sensitized lymphocytes which mediate the disease. The long-term suppression could be explained as the complete destructive action on all sensitized cells including those in central lymphoid tissue ; an alternate explanation involves participation of humoral antibody which was found in 30 percent of the animals two weeks after the last suppressive injection. The predominant role of sensitized cells in the immunopathology of EAE has been shown by many studies (1). Antilymphocytic serum has been shown effective in suppressing EAE if given soon after sensitization (53). Moreover, if lymphocytes are pretreated with neural tissue, they become ineffective in the passive transfer of EAE (63).

Significance of suppression. - This study appears to be one of the first cases of complete reversal of an experimentally produced autoimmune disease after manifestation of definite clinical signs. With the exception of cyclophosphoamide treatment of rats (64), it should be stressed that chemical agents, which inhibit EAE if given near the time of sensitization, are not capable of suppressing disease if given after onset of clinical signs. Because of its specificity, it is apparent that the use of a purified, defined antigen is preferable to immunosuppressive drug therapy. Our results, therefore, may suggest a relevant clinical approach in human immunopathology, particularly in demyelinating disease where similarities to EAE exist.

REFERENCES

1 ALVORD, E.C., in Handbook of Clinical Neurology (Ed. VINKEN, P. and BRUYN, G.) North Holland Publ. Co, Amsterdam, 1970, p.500.

2 ALVORD, E, HURBY, S., SHAW, C. and KIES, M., Int. Arch. Allergy, Suppl. 36 (1969) 203.

3 ALVORD, E., SHAW, C., HRUBY, S. and KIES, M., Ann. N.Y. Acad. Sci., 122 (1965) 333.

4 BALDWIN, G. and CARNEGIE, P., Biochem. J., 123 (1971) 69.

5 BARMAN,T. and KOSHLAND, D., J. Biol. Chem., 242 (1967) 5771.

6 BERGSTRAND, H., Eur. J. Biochem., 21 (1971) 116.

7 BOSMANN, B., HAGOPIAN, A. and EYLAR, E.H., Arch. Biochem.Biophys., 128 (1968) 51.

8 BROSTOFF, S. and EYLAR, E.H., Proc. Natl. Acad. Sci. US, 68 (1971) 769.

9 BROSTOFF, S., ROSEGAY, A. and VANDEN HEUVEL, W.J., Arch. Biochem. Biophys., submitted.

10 BURNETT, P. and EYLAR, E.H., J. Biol. Chem., 246 (1971) 3425.

11 CARNEGIE, P., Nature, 223 (1969) 958.

12 CARNEGIE, P., Biochem. J., 111 (1969) 240.

13 CARNEGIE, P., Biochem. J., 123 (1971) 57.

14 CASPARY, E. and FIELD, E.J.,Ann. N.Y. Acad. Sci., 122 (1965)182.

15 CHAO, L.P. and EINSTEIN, E., J. Neurochem., 17 (1970) 1121.

16 CHAO, L.P. and EINSTEIN, E., J. Biol. Chem., 245 (1970) 6397.

17 CUNNINGHAM, V. and FIELD, E.J., Ann. N.Y. Acad. Sci., 122 (1965) 346.

18 DANIELLI, J. and DAVSON, H., in The Permeability of Natural Membranes, Cambridge Univ. Press, New York, 1943.

19 DEHLINGER, P. and SCHIMKE, R., J. Biol. Chem., 246 (1971) 2574.

20 EINSTEIN, E., CSEJTEY, J., DAVIS, W. and RAUCH, H., Immunochem., 5 (1968) 567.

21 ENG, L., CHAO, F., GERSTL, B., PRATT, D. and TAVASTSTJERNA,M., Biochemistry, 7 (1968) 4455.

22 EYLAR, E.H., J. Theor. Biol., 10 (1965) 89.

23 EYLAR, E.H., Proc. Natl. Acad. Sci. US, 67 (1970) 1425.

24 EYLAR, E.H. in Immunological Disorders of the Nervous System (Ed. ROWLAND, L.P.) Williams and Wilkins, 1971, p.50.

25 EYLAR, E.H., BRODY, O., MADOFF, M. and ONCLEY, J.L., J. Biol. Chem., 237 (1962) 1992.

26 EYLAR, E.H. and BROSTOFF, S., J. Biol. Chem., 246 (1971) 3418.

27 EYLAR, E.H., BROSTOFF, S., HASHIM, G., CACCAM, J. and BURNETT, P., J. Biol. Chem., 246 (1971) 5770.

28 EYLAR, E.H., BROSTOFF, S., JACKSON, J. and CARTER, H., Proc. Natl. Acad. Sci. US, 69 (1972) 617.

29 EYLAR, E.H., CACCAM, J. and JACKSON, J., Trans. Am. Soc. Neurochem., 1970.

30 EYLAR, E.H., CACCAM, J., JACKSON, J. and ROBINSON, A., Science, 168 (1970) 1220.

31 EYLAR, E.H. and HASHIM, G., Proc. Natl. Acad. Sci. US, 61 (1968) 644.

32 EYLAR, E.H. and HASHIM,G. , Arch. Biochem. Biophys., 131 (1969) 215.

33 EYLAR, E.H., JACKSON, J., ROTHENBERG, B. and BROSTOFF, S., Nature, in press.

34 EYLAR, E.H., SALK, J., BEVERIDGE, G. and BROWN, L., Arch. Biochem. Biophys., 132 (1969) 34.

35 EYLAR, E.H. and THOMPSON, M., Arch. Biochem. Biophys., 129 (1969) 568.
36 FINEAN, J., Exptl. Cell Res., 5 (1953) 202.
37 GREEN, D.E. and PERDUE, J., Proc. Natl. Acad. Sci. US, 55 (1966) 1295.
38 HAGOPIAN, A. and EYLAR, E.H., Arch. Biochem. Biophys., 126 (1968) 785.
39 HAGOPIAN, A. and EYLAR, E.H., Arch. Biochem. Biophys., 129 (1969) 515.
40 HAGOPIAN, A. and WHITEHEAD, J., Feder. Proc., 30 (1971) 1186.
41 HAGOPIAN, A., WHITEHEAD, J., WESTALL, F. and EYLAR, E.H., J. Biol. Chem., 246 (1971) 2519.
42 HALPERIN, A., BRANER, A. and ALEXANDER, E., Phys. Rev., 108 (1957) 928.
43 HASHIM, G. and EYLAR, E.H., Arch. Biochem. Biophys., 135 (1969) 324.
44 HASHIM, G. and EYLAR, E.H., Arch. Biochem. Biophys., 129 (1969) 635.
45 KABAT, E., WOLF, A. and BEZER, A., J. Exptl. Med., 85 (1947) 117.
46 KIBLER, R., McKNEALLY, S., SELDEN, P. and CHOW, F., Science, 164 (1969) 577.
47 KIBLER, R. and SHAPIRA, R., J. Biol. Chem., 243 (1968) 281.
48 KIES, M., Ann. N.Y. Acad. Sci., 122 (1965) 161.
49 KIES, M., THOMPSON, E. and ALVORD, E., Ann. N.Y. Acad. Sci., 122 (1965) 122.
50 KORN, E., Science, 153 (1966) 1491.
51 LAMPERT, P., Acta Neuropath., 9 (1967) 99.
52 LENNON, V., WILKS, A. and CARNEGIE, P., J. Immunol., 105 (1970) 1223.
53 LIEBOWITZ, S., LESSOF, M. and KENNEDY, L., Clin. Exptl. Immunol., 3 (1968) 753.
54 LUMSDEN, C., ROBERTSON, D. and BLIGHT, R., J. Neurochem., 13 (1966) 127.
55 MANDEL, P., Abstr. Third Intern. Meet. Intern. Soc. Neurochem., 1971, p.418.
56 MARTENSON, R., DIEBLER, G. and KIES, M., J. Biol. Chem., 244 (1969) 4261.
57 MARTENSON, R. and GAITONDE, M., J. Neurochem., 16 (1969) 889.
58 NAKAO, A., DAVIS, W. and EINSTEIN, E., Biochim. Biophys. Acta, 130 (1966) 163.
59 NAKAO, A., DAVIS, W. and EINSTEIN, E., Biochim. Biophys. Acta, 130 (1966) 171.
60 OSHIRO, Y. and EYLAR, E.H., Arch. Biochem. Biophys., 138 (1970) 392.
61 PAIK, W. and KIMA, S., J. Biol. Chem., 245 (1970) 88.
62 PALMER, F. and DAWSON, R., Biochem. J., 111 (1969) 629.
63 PATERSON, P., in International Convocation on Immunology (Ed. ROSE, N. and MILGROM, F.) S. Karger, New York, 1969, p.260.
64 PATERSON, P. and DROBISH, D., Science, 165 (1969) 191.

65 ROBERTSON, D., BLIGHT, R. and LUMSDEN, C., Nature, 196 (1962) 1005.

66 ROBOZ-EINSTEIN, E., ROBERTSON, D., DI CAPRIO, J. and MOORE, W., J. Neurochem., 9 (1962) 353.

67 SEIL, F., FALK, G., KIES, M. and ALVORD, E., Exptl. Neurol., 22 (1968) 545.

68 SPITLER, L., VON MULLER, C., FUDENBERG, H. and EYLAR, E.H., J. Exptl. Med., in press.

69 SMYTH, D. and UTSUMI, S., Nature, 216 (1967) 332.

70 WESTALL, F., ROBINSON, A., CACCAM, J., JACKSON, J. and EYLAR, E.H., Nature, 168 (1970) 1220.

THE CONTROL OF MYELIN SYNTHESIS AND INBORN ERRORS OF METABOLISM

P. MANDEL

Centre de Neurochimie du CNRS, 67-Strasbourg (France)

Morphological evidence indicates that glial cells proliferate and differentiate shortly before myelination. Myelin is then formed in the central nervous system by wrapping of oligodendrocyte plasma membrane around the axon (4, 15, 16). Thus in the biogenesis of myelin, it is necessary to consider the phenomenon at three levels; a morphological level, a supramolecular level and a molecular level.

At the morphological level we have to take into account phenomena associated with the proliferation, migration and differentiation of glial cells, involving in particular an increase in cellular mass and the arrangement of the oligodendroglia in chains.

At the supramolecular level we must study the elongation of glial cell membranes, and the processes of controlled addition of components to, and loss of components from these proliferating membranes.

At the molecular level it is necessary to understand the synthesis of myelin-specific proteins such as the proteolipids (7),the basic proteins (18,19), the "Wolfgram" proteins (34) and possibly other proteins. It is also necessary to study the biosynthesis of myelin-specific lipids such as the galactolipids with long-chain unsaturated and hydroxy fatty acids. It is necessary, however, to bear in mind that these lipids may exist in low concentrations in other cells. Finally, it is necessary to analyse the biosynthesis of lipids which are not specific to myelin such as the phospholipids and cholesterol. The control of the molecular processes may occur by feedback inhibition of rate-limiting enzymes, by substrate regulation or by induction, and repression at the genetic level, of both the synthetic and degradative pathways.

Myelin Proteins

Given that the genetic code and the mechanism of protein biosynthesis are universal, it must be accepted that the deficiency in the biosynthesis of myelin proteins will occur either at the transcription level by regulation of the synthesis of mRNA or at the translation level on polysomes, due to a defect on the general protein biosynthetic system. Many studies have shown that deposition of myelin lipids starts in the very young animal (around 10 days in rat and mouse). This process then reaches a maximum rate around 19 days, decreasing thereafter. Since proteins form part of the basic skeleton of myelin, one would expect parallel changes in the biosynthesis of proteins, that is either in the rate of transcription of the corresponding mRNA's or in that of the translation of mRNA's during myelination. In view of the difficulty of isolating and characterizing specific mRNA's, this hypothesis can only be tested in an indirect way. Thus, we investigated the change in myelin proteolipids during ontogenesis in normal mice and in the myelin-deficient mice described by Sidman, Dickie and Appel (30).

Proteolipids seem to be the only protein fraction which can be quantitatively studied in total brain. The lack of the proteolipid band in the electrophoregrams of homogenates of myelin-deficient mutant brains suggests strongly that other protein bands do not overlap with the myelin proteolipid band (Fig. 1). During myelination, the amount of proteolipids in control mouse brains (estimated by densitometry of protein profiles in polyacrylamide gels) increased from 18 to 772 μg per brain. In the Jimpy mutants, the level of 35 μg found at 15 days remained constant or diminished slightly. In the Quaking mutants, there was less than 10 % of the normal amount of the myelin proteolipid (Table I).

TABLE I

Absolute Levels of Band 7 and Band 11 Proteolipids in Normal and Jimpy Mice as a Function of Age

Age	Normal		Jimpy		Normal		Quaking	
	7	11	7	11	7	11	7	11
10 days	18.4	101.5						
15 days	127	150	35.5	175				
25 days	395	132	30.2	196				
Adult	772	200			830	162	67.5	290

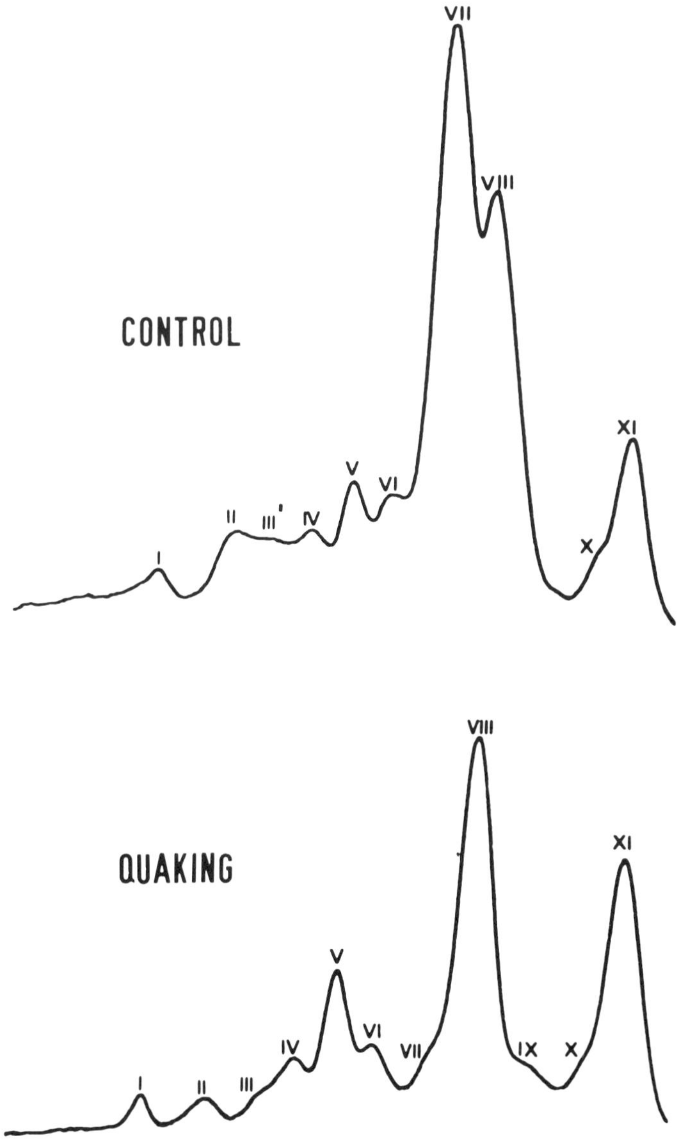

Fig. 1. Densitometric profiles of electrophoretograms of proteolipid proteins extracted from control and Quaking mutant mouse brains.

The choice between deficiency in transcription or in translation cannot be made without further experimental evidence. However, the decrease in the proteolipid protein levels in Quaking, without an equally marked decrease in basic protein levels found also by Jacque _et al_. (9), is rather in favour of a defect at the transcription level. This tends to rule out a general deficiency in the protein biosynthesis machinery at the ribosomal level.

2',3'-Cyclic AMP 3'-Phosphohydrolase

This phosphohydrolase described by Drummond, Iyer and Keith (5) was found by Kurihara and Tsukada (13,14) to be concentrated in the myelin fraction. There was a rapid accumulation of this enzyme activity in normal mice, whereas this enzyme is present at much lower levels in the Jimpy and Quaking mice (11,12). Quaking mice show histologically more myelin than the Jimpy (6,30,35) and there was more enzyme activity in the Quaking mice than in the Jimpy. Since this enzyme does not decrease in activity after myelination, as is the case for the enzymes involved in myelin biosynthesis,this enzyme can be used as a marker for myelin.

Gangliosides

Various workers have shown that one of the most striking differences between normal and myelin-deficient brains is the markedly reduced level of cerebrosides and sulphatides in the latter (2, 8, 26, 27, 30). Preliminary studies (26) suggested that ganglioside levels were hardly affected, thus suggesting that there was no defect in ceramide synthesis in the mutants.

However, a closer examination of the results has shown that the ganglioside GM_1 was present in lower amounts in the mutant mice (Table II). This can be related to the work (31, 32) which suggests strongly that at least a part of this ganglioside is found in myelin. But since the reduction is much less marked for this ganglioside than for the cerebrosides, it suggests that this ganglioside is neither exclusively localized in myelin, nor produced exclusively by oligodendroglia. Therefore our initial conclusion that ceramide synthesis is not blocked in the neurological mutants must be qualified to allow for a possible block of ceramide synthesis in oligodendroglia.

Studies of the UDP-glucose : ceramide glucosyl transferase activity as a function of age have shown for the normal mouse a maximum enzyme activity before myelination followed by a decrease in the enzymatic activity (Fig.2). In the mutants the values are very close to normal at all times.

TABLE II

Ganglioside Distribution in Control, Quaking and Jimpy Mouse Brains

	28 days			60 days		
	Control		Jimpy	Control		Quaking
G_5 (GM_2)	4.6		4.2	14.3	+	10.2
G_4 (GM_1)	12.1	+	8.0			
G_{3a}(GD_3)	7.2		7.9	38.1		37.1
G_3 (GD_{1a})	30.6		30.3			
G_{2a}(GD_2)	10.2		11.4	10.5		10.4
G_2 (GD_{1b})	12.6		12.6	12.8		12.9
G_1 (GT_1)	15.3		15.4	16.3		14.9
G_U (GQ_1)	4.9		4.6	4.9		5.0
Origin	2.7		3.4	3.6		4.1

+ $p < 0.01$. The results (average of 4-6 experiments are expressed as a percent of total NANA. The ganglioside nomenclature of Korey and Gonatas (8) is used, with Svennerholm (30) equivalents in brackets.

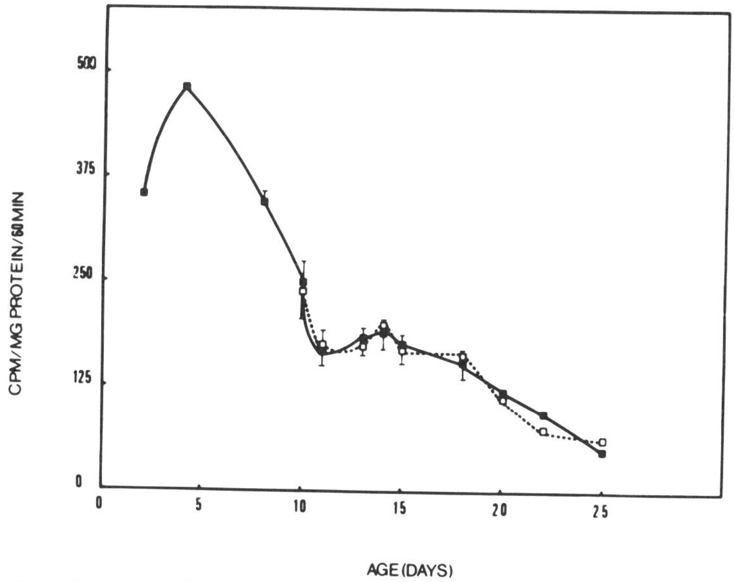

Fig. 2. Developmental curve of UDP-glucose ceramide glucosyl transferase with age in control —— and mutant ---- brains.

Cerebrosides

The pathway for cerebroside synthesis can pass through sphingosine or ceramide (Fig. 3). In normal mice UDP-galactose :sphingosine galactosyl transferase appeared at high levels during myelination but later decreased in activity (20,22). The same pattern was observed in the Jimpy and Quaking mice, although the levels of enzyme activity reached were very much lower. We have ruled out the possibility of activation and inhibition effects, rather than of enzyme synthesis, by performing crossed incubation both for extracts from normal mice at different ages, and from Jimpy mice. The values obtained were close to those expected assuming that there was no inhibitory or activating effect. The parallel development of this enzyme and the myelin defect in the mutant myelin-deficient mice is an indication that this metabolic pathway exists *in vivo*.

For the UDP-galactose : ceramide galactosyl transferase the evolution of the enzymatic activity as a function of age was similar to that of UDP-galactose sphingosine galactosyl transferase : increased enzyme activity during myelination followed by a decreased enzyme activity. Thus it seems likely that this pathway also exists *in vivo*. In the mutants a considerable reduction in the enzyme activity was observed (21). Inhibitory effects were also ruled out in this case. These enzyme activities were however present in the peripheral nervous system of the mutant mice where myelination is not affected.

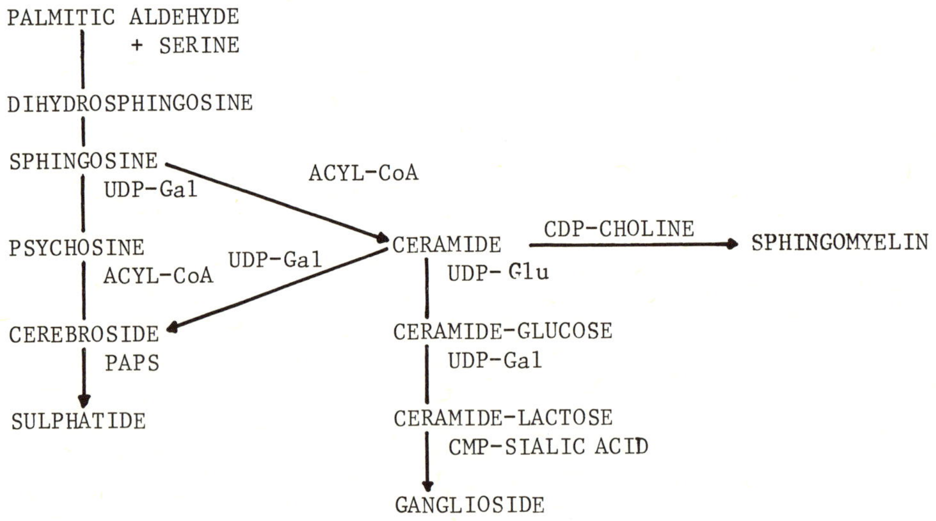

Fig. 3. Possible pathways for sphingolipid biosynthesis in brain.

Sulphatides

McKhann and Ho (17) have shown that the enzyme activity for the incorporation of sulphate either in a simple homogenate or in a homogenate containing galactolipids, reaches a maximum during myelination. In normal mice we have studied two possibilities : the transfer of sulphate to psychosine or to cerebrosides. The activity of transfer of sulphate to psychosine developed in a bell-shaped curve (24) as did the transfer of sulphate from PAPS to cerebrosides (29). The only other organ in which this enzyme for the synthesis of sulphatides is found in a high amount is the kidney. In the mutant, in the premyelination period, the values for this enzyme activity were close to normal values in both organs, kidney and brain. During the period of myelination, the values in the mutant brain were lower than the normal, but the kidney enzyme was not affected (Fig. 4). It is

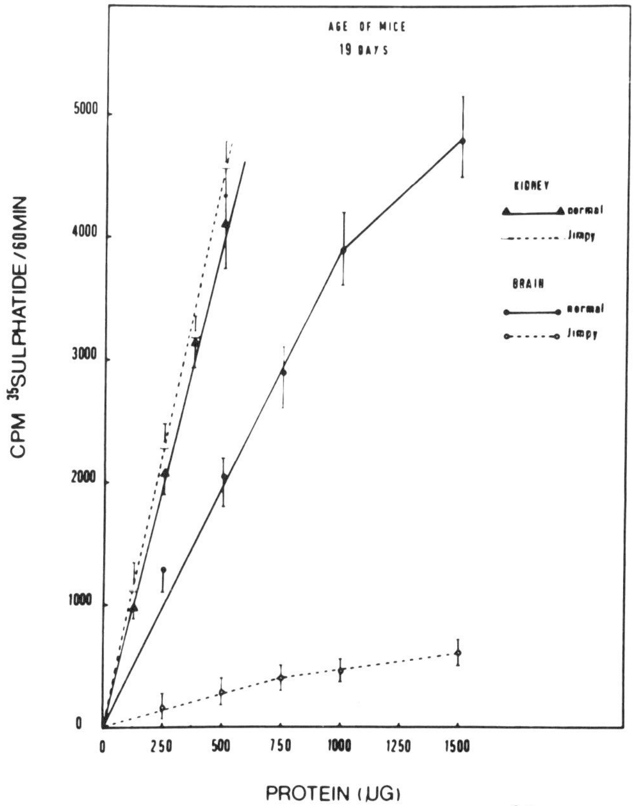

Fig. 4. *In vitro* formation of sulphatides from ^{35}S-PAP in the presence of exogenous galactocerebrosides by the PAPS-cerebroside sulphotransferase of normal and Jimpy mice as a function of enzyme concentration. Each point is the average of 3 experiments. Vertical bars represent the range of values.

possible that this type of curve in the mutant mice can be interpreted in terms of two enzymes. One enzyme concerned with the synthesis of extramyelin sulphatides which is not affected,whereas the enzyme involved in myelin sulphatide synthesis, presumably found in the oligodendrocyte, is reduced. The lack of sulphatide is not due to the cerebroside deficiency since we tested sulphatide synthesis in the presence of exogenous cerebrosides (29).

Long-Chain Fatty Acids

It is known that cerebrosides contain long-chain unsaturated and hydroxy fatty acids. A reduction in the levels of these fatty acids has been observed in Quaking mice (2) and we have observed a similar reduction in Jimpy mice (25). The pathway of fatty acid elongation seems disturbed (3, 28).

Discussion

Several questions concerning the enzyme deficiency of myelin formation can be discussed in the light of the data obtained in the myelin deficient mutants. The "premyelin" fraction observed by Banik and Davison (1) is relatively poor in cerebrosides and sulphatides but contains 2,3'-cyclic AMP 3'-phosphohydrolase. This fraction may exist in the mutant. The decrease in brain proteins, after cerebroside synthesis should have commenced, may correspond to a degradation of the premyelin fraction. However membranes other than myelin containing 2',3'-cyclic AMP 3'-phosphohydrolase have been reported (36). It could be that the "premyelin" is a subellular fraction somewhat enriched in glial cell plasma membrane upon which the assembly of myelin specific lipids starts.

Summarizing the data on the mutant mice, we see that there was a lack, not only in the synthesis of long chain unsaturated and hydroxy fatty acids as was found in Quaking mice (2,3,28) but of at least four other enzymes. The synthesis of myelin-specific proteins was also affected. This indicates that the biogenesis of myelin is a coordinated phenomenon as studies on normal mice would suggest, where a whole series of enzymes is transcribed and translated in parallel. Some hypotheses for the regulation of this process can be envisaged :

- There could be induction of a "myelin-enzyme" operon.
- There could be a sequential transcription phenomenon where one enzyme or its products regulate the transcription of enzymes involved in later stages of myelin biosynthesis.

Both these processes could be dependent upon prior induction of oligodendrocyte migration and differentiation. Thus the basic defect of the mutant could be related to an impairement of the differentiation of oligodendrocytes prior to the synthesis of the enzymes directly responsible for the synthesis of myelin constituents.

Thus studies on the mutant mice have clarified a certain number of aspects of myelin biosynthesis and the metabolic pathways involved. There is a segregation of genetic control of myelination in the CNS and in peripheral nerves. The transcription and translation of enzymes involved in the biosynthesis of myelin specific lipids are coordinated phenomena, induced at a precise period of ontogenesis. After the myelination crisis these enzyme activities decrease. Oligodendroglial differentiation and myelination seem to be genetically controlled by at least two different chromosomes, since the Jimpy mutation is sex-linked while Quaking mutation is autosomal recessive.

REFERENCES

1 BANIK, M.L. and DAVISON, A.N., Biochem. J., 115 (1969) 1051.
2 BAUMANN, N.A., JACQUE, C.M., POLLET, S. and HARPIN, M.L., Europ. J. Biochem., 4 (1968) 340.
3 BOURRE, J.M., POLLET, S., DAUDU, O. and BAUMANN, N., C.R. Acad. Sci. (Paris), Sér.D, 273 (1971) 1534.
4 BUNGE, R.P., Physiol. Rev., 48 (1968) 197.
5 DRUMMOND, G.I., IYER, N.T. and KEITH, J., J. Biol. Chem., 237 (1962) 3535.
6 FARKAS, D., ZAHND, J.P., NUSSBAUM, J.L. and MANDEL, P., in Les Mutants Pathologiques chez l'Animal (Ed. SABOURDIE, M.) CNRS, Paris, 1970, p.21.
7 FOLCH, J. and LEES, M., J. Biol. Chem., 191 (1951) 807.
8 GALLI, C. and RE CECCONI GALLI, D., Nature (Lond.), 220 (1968) 165.
9 JACQUE, C.M., LOUIS, C.F., GUEDE, C. and BAUMANN, N., C.R. Acad. Sci. (Paris), Sér.D , 274 (1972) 126.
10 KOREY, S.R. and GONATAS, J., Life Sci., 2 (1963) 296.
11 KURIHARA, T., NUSSBAUM, J.L. and MANDEL, P., Brain Res., 13 (1969) 401.
12 KURIHARA, T., NUSSBAUM, J.L. and MANDEL, P., J. Neurochem., 17 (1970) 993.
13 KURIHARA, T. and TSUKADA, Y., J. Neurochem., 14 (1967) 1167.
14 KURIHARA, T. and TSUKADA, Y., J. Neurochem., 15 (1968) 827.
15 LUSE, S., J. Biophys. Biochem. Cytol., 2 (1956) 777.
16 LUSE, S., J. Biophys. Biochem. Cytol., 2 (1956) 531
17 McKHANN, G.M. and HO, W., J. Neurochem., 14 (1967) 717.
18 NAKAO, A., DAVIS, W.J. and EINSTEIN, E.R., Biochim. Biophys. Acta, 130 (1966a) 163.
19 NAKAO, A., DAVIS, W.J. and EINSTEIN, E.R., Biochim. Biophys. Acta, 130 (1966) 171.
20 NESKOVIC, N.M., NUSSBAUM, J.L. and MANDEL, P., Febs Letters, 3 (1969) 199.

21 NESKOVIC, N.M., NUSSBAUM, J.L. and MANDEL, P., Febs Letters, 8 (1970) 213.
22 NESKOVIC, N.M., NUSSBAUM, J.L. and MANDEL, P., Brain Res., 21 (1970) 39.
23 NESKOVIC, N.M., SARLIEVE, L.L. and MANDEL, P., Brain Res., 42 (1972) 147.
24 NUSSBAUM, J.L. and MANDEL, P., J. Neurochem., 19 (1972) 1789.
25 NUSSBAUM, J.L., NESKOVIC, N.M., KOSTIC, D.M. and MANDEL, P., Bull. Soc. Chim. Biol., 50 (1968) 2194.
26 NUSSBAUM, J.L., NESKOVIC, N.M. and MANDEL, P., J. Neurochem., 16 (1969) 927.
27 NUSSBAUM, J.L., NESKOVIC, N.M. and MANDEL, P., J. Neurochem., 18 (1971) 1529.
28 POLLET, S., BOURRE, J.M. and BAUMANN, N., C.R.Acad.Sci.(Paris), 268 (1969)2146.
29 SARLIEVE, L.L., NESKOVIC, N.M. and MANDEL, P., Febs Letters, 19 (1971) 91.
30 SIDMAN, R.L., DICKIE, M.M. and APPEL, S.H., Science, 144 (1964) 309.
31 SUZUKI, K., J. Neurochem., 17 (1970) 209.
32 SUZUKI, K., PODUSLO, S.E. and NORTON, W.T., Biochim. Biophys. Acta, 144 (1967) 375.
33 SVENNERHOLM, L., J. Neurochem., 10 (1963) 613.
34 WOLFGRAM, F., J. Neurochem., 13 (1966) 461.
35 ZAHND, J.P. and BONAVENTURE, N., C.R.Soc. Biol., 163 (1969) 1631.
36 ZANETTA, J.P., BENDA, P., GOMBOS, G. and MORGAN, I.G., J.Neurochem., 19 (1972) 881.

ISOLATION AND CHARACTERIZATION OF MYELIN PROTEIN FROM ADULT QUAKING MICE AND ITS SIMILARITY TO MYELIN PROTEIN OF YOUNG NORMAL MICE

Pierre MORELL, Seymour GREENFIELD, William T. NORTON and Henry WISNIEWSKI
Saul R. Korey Department of Neurology and Department of Pathology (Neuropathology), Albert Einstein College of Medicine, Bronx, New York 10461 (USA)

Quaking is a genetically determined neurological disorder of mice in which the homozygotes are characterized by marked tremors and seizures. Sidman, Dickie and Appel (16) demonstrated histologically that the CNS showed incomplete myelination without marked changes in the axons. Ultrastructural studies (4, 18) have shown that the myelin sheaths are thin and uncompacted, that there is uneven growth of lateral loops, and that nodes of Ranvier are poorly developed (Fig. 1). These observations regarding the CNS of the mature Quaking mouse are very similar to those seen in the normal mouse during the early period of myelinogenesis (8-12 days).

Lipid analyses by several laboratories (3, 5, 9, 15) indicate low levels of myelin specific lipids in Quaking, as compared to control mice. The depression of galactosylceramide (cerebroside) levels is especially marked. An analogy between the lipid composition of myelin of adult Quaking mice and young controls has been demonstrated (17). We have investigated the terminal step in biosynthesis of galactosylceramide, galactosylation of ceramide by UDP-galactose : ceramide galactosyl transferase (6). *In vitro* this enzymatic activity is about 60 percent depressed in Quaking mice during the age period of most rapid myelination. Related enzymatic activities studied as controls included condensation of palmitoyl-CoA and serine to form ketodihydrosphingosine, acylation of sphingosine to form ceramide and glucosylation of ceramide to form glycosyl-ceramide. These enzymatic activities were present at the same level in mutant mice and normal littermates. Independent observations indicating a depression of cerebroside biosynthesis in Quaking have been made by Neskovic, Nussbaum and Mandel (12).

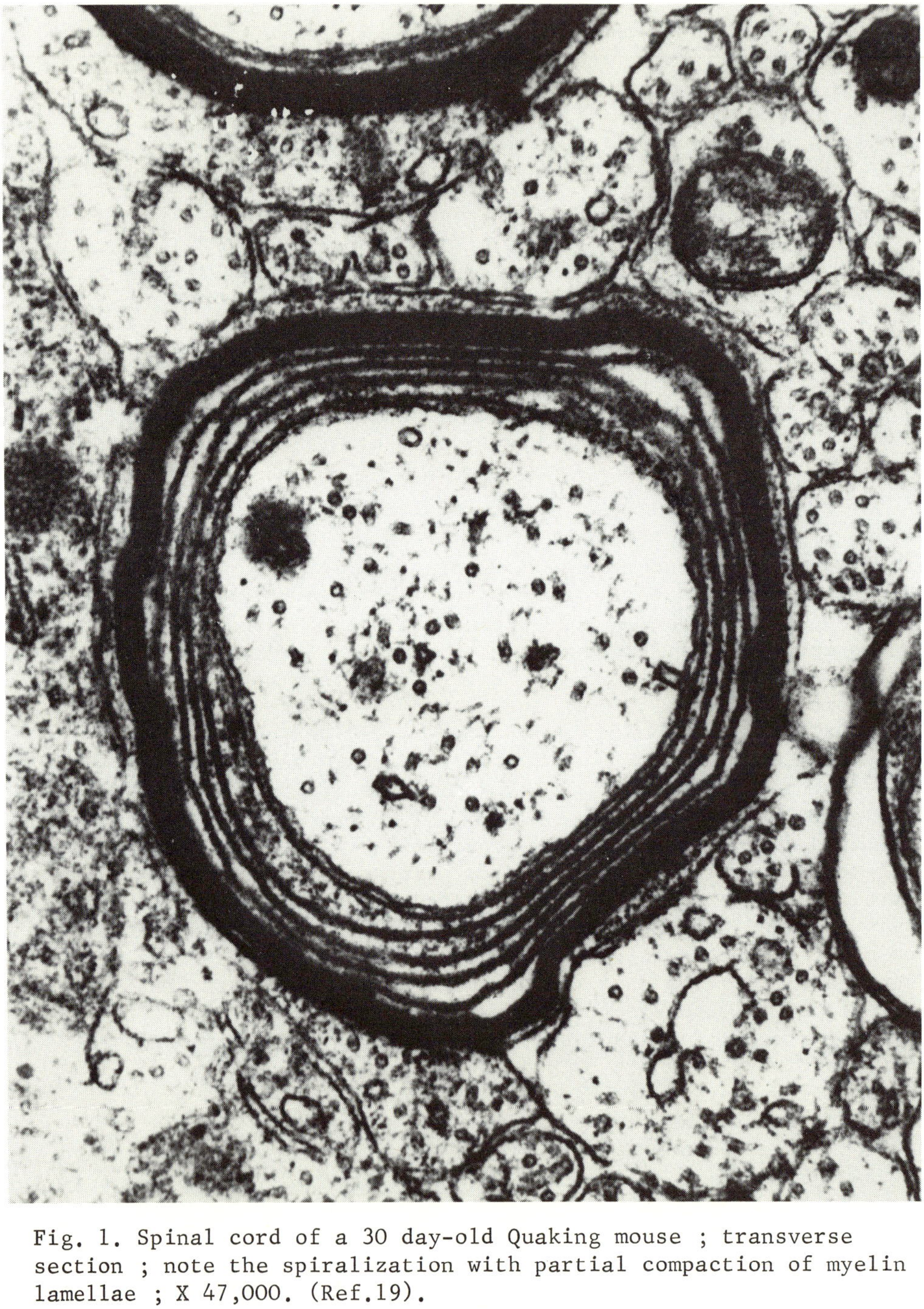

Fig. 1. Spinal cord of a 30 day-old Quaking mouse ; transverse section ; note the spiralization with partial compaction of myelin lamellae ; X 47,000. (Ref.19).

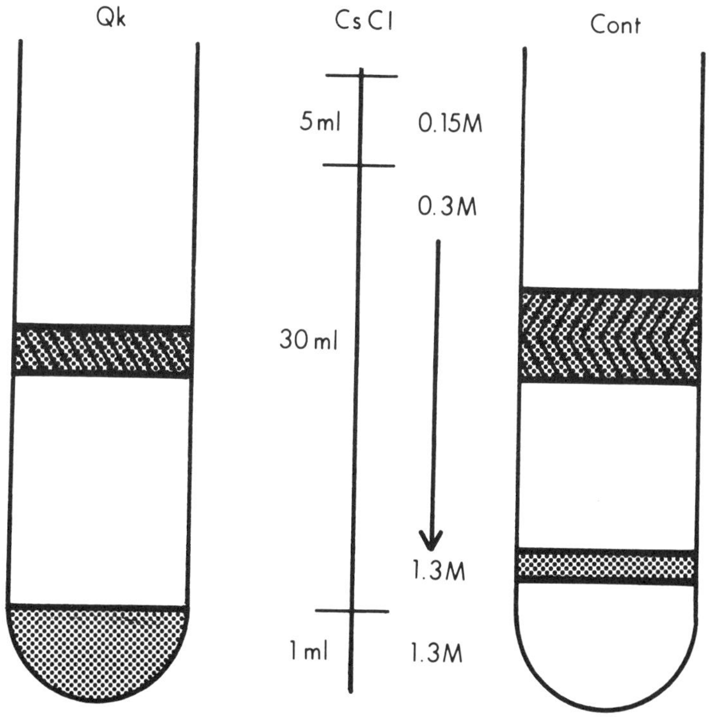

Fig. 2. Schematic representation of the distribution of material after isopycnic separation of crude myelin on a continuous CsCl gradient. Results with preparations from Quaking (Qk) and controls (Cont) mice are shown.

Since the morphological observations and studies of brain lipids indicated that the mutation was expressed primarily as an aberration involving myelin,we decided to investigate the protein composition of myelin. In order to establish the quantitative relationship of various minor protein components to each other, and to the major proteins, it was necessary to study isolated myelin. The method used to isolate myelin, was that of Norton and Davison (in preparation ; see also Ref.13) and involves preliminary separation from other subcellular fractions on a discontinuous sucrose gradient. The crude myelin thus obtained is subjected to osmotic shock and further purified on a CsCl gradient. The CsCl gradient (Fig. 2) separates the crude myelin of normal mice into two well separated bands. Often a partial separation of the upper band into two components is observed if the gradients are handled very carefully. The relative proportion of material in these two bands is

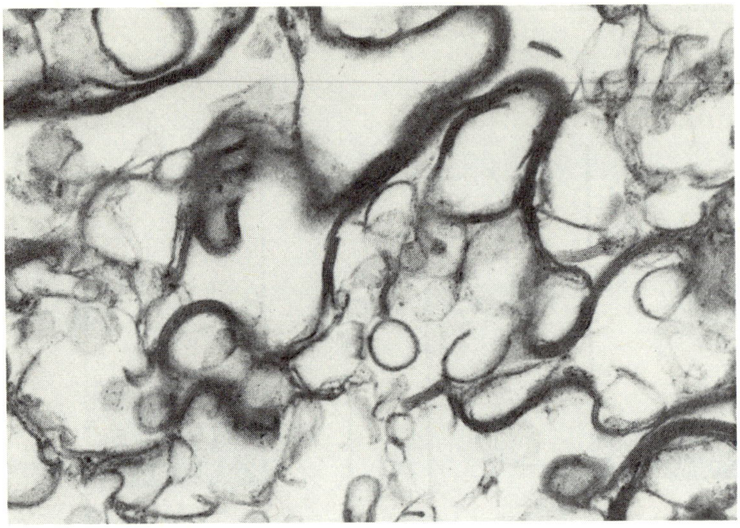

Fig. 3. "Upper layer" material from a CsCl gradient as shown in Fig. 2 (right hand tube). The crude myelin loaded on the gradient was from normal 45 day-old mice. Material from the upper layer was removed with a pipette, diluted with water, collected by centrifugation and prepared for electron microscopy as described by Raine, Poduslo and Norton (14). The only identifiable elements are myelin lamellae ; X 52,000. (Reduced 40% for reproduction.)

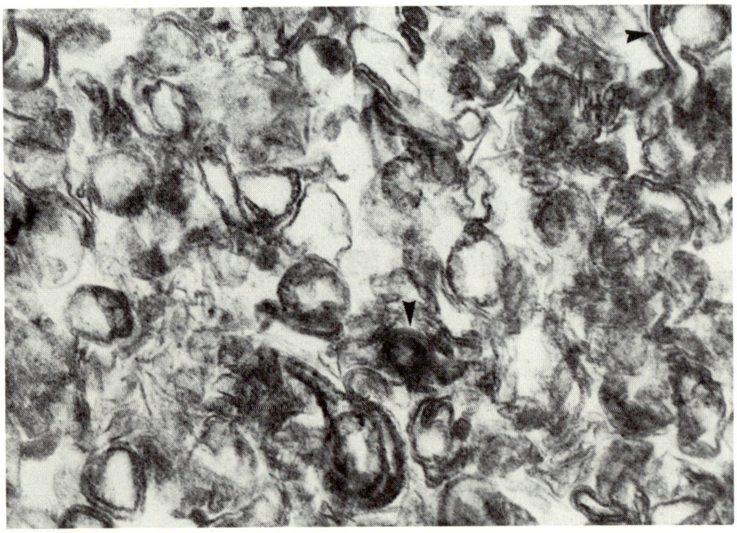

Fig. 4. Material from the lower layer of a CsCl gradient (same preparation as described in Fig.3) was examined. Between the aggregates of unidentifiable membranous material,scattered profiles of short segments of myelin lamellae (arrows) are visible;X 65,000. (Reduced 40% for reproduction.)

age dependent (see below) but all the material in the upper band appears to be highly purified myelin by electron microscopic criteria (Fig.3). The lower layer of material (Fig.4) contains many small pieces of myelin, and large amounts of unidentifiable membranous material. In some sections a significant percentage of material identifiable as the membranous components of endoplasmic reticulum and mitochondria was visible. This material is probably the "myelin-like" fraction described by Banik and Davison (2) and Agrawal et al. (1). The phrase "myelin-like" as used by these authors is based on certain biochemical similarities and does not refer to ultrastructure. Although recognizing the priority of these authors in studies of such a fraction we feel that the term "myelin-like" may eventually lead to confusion, since myelin is defined as a morphological entity. The subcellular fraction under consideration, as characterized ultrastructurally both by Agrawal et al. (1) and ourselves, does not morphologically resemble myelin but is a membranous fraction of unique density. We will therefore preferentially use the operational term "lower layer material".

When myelin is isolated from the brains of Quaking mice by the CsCl technique a band of myelin formed in the CsCl gradient. This band was slightly lower than the myelin from control mice, and, by visual inspection, appeared to correspond in density to the lower part of the upper band from control preparations (Fig. 2). The amount of myelin isolated from Quaking mice was only 5-10 percent of that from littermate controls, and it could not be differentiated by electron microscopy from myelin of control mice. The gradient on which the crude Quaking myelin was separated contained a diffuse, uncompacted, pellet which was distinctly lower than the lower layer from control animals (Fig. 2). This material is electron dense and granular but has no recognizable cellular elements (Fig. 5).

To investigate the protein composition of these myelin preparations we adapted the technique of discontinuous, polyacrylamide gel electrophoresis in sodium lauryl sulfate buffers developed by Maizel and Laemmli (in preparation, see also Ref.10). In this system, separation of proteins is almost exclusively on the basis of molecular weight, the smaller proteins migrating more rapidly. Myelin from mutant and littermate control mice was electrophoresed with standards of mouse basic and proteolipid protein (Fig. 6). Myelin from bovine white matter, along with appropriate standards was also examined since mouse myelin has not previously been characterized and compared to other myelins. Standards of basic and proteolipid protein were prepared by a modification (8) of the procedure of Gonzalez-Sastre (7). The beef myelin consists primarily of basic protein and proteolipid protein in agreement with reports from many laboratories. The presence of other protein is indicated by other dye absorbing bands, two between the basic and proteolipid protein and several in the high molecular weight region of the gel.

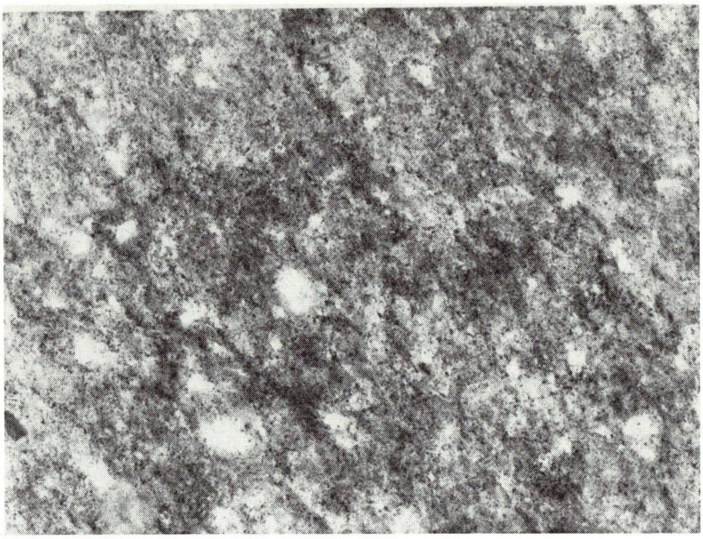

Fig. 5. Pellet fraction from a CsCl gradient as shown in Fig. 2 (left hand side). Material from this fraction was prepared for electron microscopy (legend to Fig. 3). It consists of electron dense, granular material without recognizable cytoplasmic organelles ; X 46,000. (Reduced 40% for reproduction.)

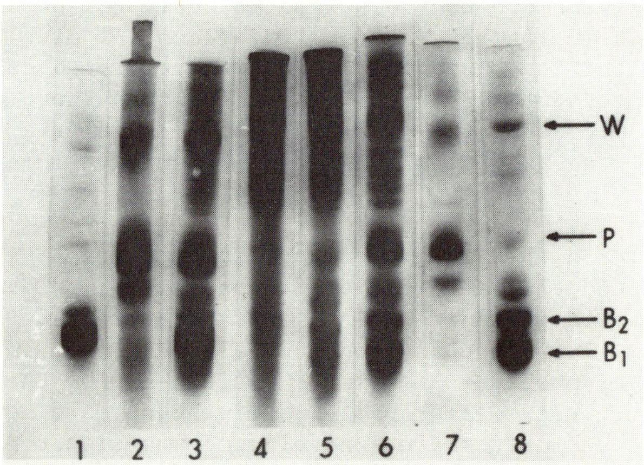

Fig. 6. Polyacrylamide gel electrophoresis of myelin proteins and standards. 150 μg of delipidated myelin protein from (3) beef,(4) 10 day-old normal mouse,(5) Quaking mouse,(6) littermate control of Quaking mouse were electrophoresed and stained with fast green. Standards (30-50 μg) include (1) beef basic protein,(2) beef proteolipid protein,(7) mouse proteolipid protein,(8) mouse basic protein. Mouse proteins are labelled B_1, (fast migrating basic protein) ; B_2,(slower migrating basic protein) ; P, (proteolipid protein); W, (major "Wolfgram" protein). (Ref.8).

The major high molecular weight protein is presumably the acidic protein described by Wolfgram and Kotorii (19). Differences between beef myelin and myelin from normal mice (Fig.6, lanes 3 and 6) include the presence of two basic proteins in the mouse sample (a result obtained with rat myelin by Martenson, Diebler and Kies,Ref. 11) and a smaller percentage of high molecular weight components in the beef myelin. When the protein pattern of myelin from Quaking mice (Fig. 6,lane 5) was compared to that from normal littermates, it was observed that although the patterns were qualitatively identical, there were major quantitative differences in the distribution of proteins. Resolution has been lost in reproduction but upon visual inspection of the original gels there are at least ten bands in the high molecular weight region and these correspond in the two gels. Myelin from Quaking mice (Fig. 6, lane 5) has much less basic protein and proteolipid protein, and consequently a much higher percentage of high molecular weight proteins. As mentioned previously the in vivo morphology of the myelin sheath was similar to that of young, 8-12 day-old animals. Therefore, we prepared myelin from 10

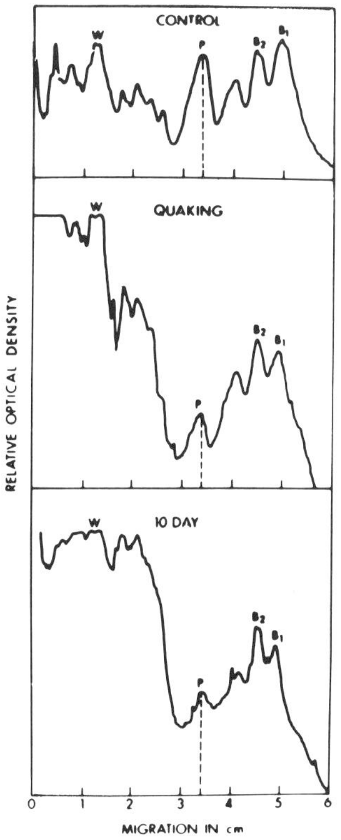

Fig. 7. Densitometry of polyacrylamide gels containing myelin proteins. Although each gel had 150 μg of protein, areas under the different tracings are not directly comparable since densitometer sensitivity was individually adjusted for each gel. (From Ref. 8).

day-old animals and ran it along with the other standards (Fig.6, lane 4). The similarity of the protein patterns of myelin obtained from Quaking animals or from young mice, in contrast to the protein distribution of myelin from adult mice, is clear in the photographs of the gels (Fig. 6). A quantitative evaluation can be made from the densitometer tracings shown in Fig. 7.

The evidence discussed above indicated that the Quaking gene is expressed largely as an arrest in myelinogenesis and therefore we started to examine the pattern of myelin maturation in control mice more carefully. As indicated in Fig. 2, crude myelin from normal mice separates into two bands on a CsCl gradient. The amount of material in each band is dependent on the age of the animals and such data is plotted in Fig. 8. The amount of lower layer material remains relatively constant with age, while mature myelin accumulates in the upper band. The relationship of the myelin and "lower layer" fractions was studied by comparing the protein composition of each fraction at various age points (Fig. 9) ; as can be seen all samples contain the same protein bands in varying proportions. The myelin proteins of mature mice consist largely of basic and proteolipid protein although a significant percentage of high molecular

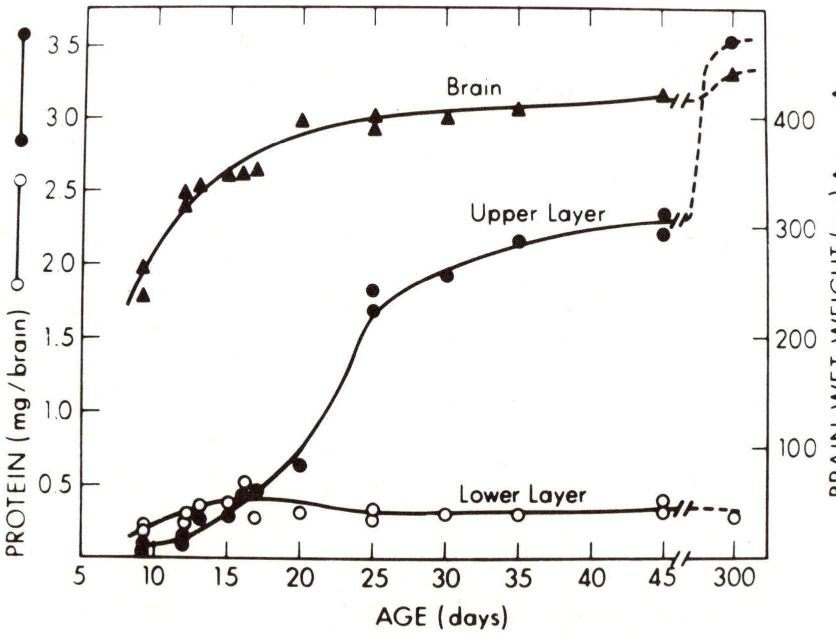

Fig. 8. Crude myelin, isolated from the brains of mice of varying ages, was fractionated on a CsCl gradient (see Fig. 2). Protein in the upper and lower layers of the CsCl gradient was determined, as well as original wet weight of brain.

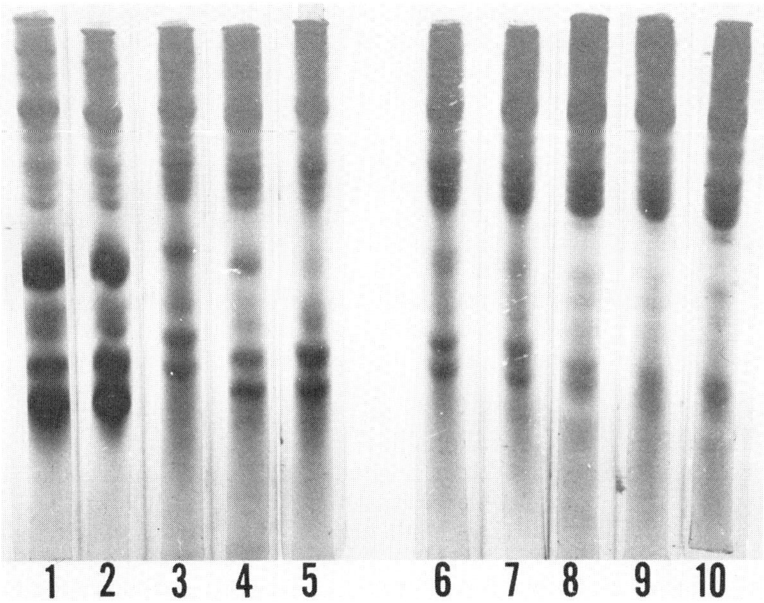

Fig. 9. Polyacrylamide gel electrophoresis of myelin and "lower layer" fractions prepared from the brains of mice of various ages. Gels (1-5) are upper layer (myelin): (1), (300 days) ; (2), 30 days; (3), 11 days ; (4), 9 days ; (5), 42 day-old Quaking mice. Gels 6-9 lower layer (myelin-like) : (6), 300 days ; (7), 30 days ; (8),11 days ; (9), 9 days. (10) is the pellet fraction from 42 day-old Quaking mice.

weight proteins are also present. There is a gradient of protein distribution with age, the younger the animal the less proteolipid and basic protein is present relative to the high molecular weight protein. Densitometry indicates that the younger animals, and adult Quaking mice, have half as much basic protein, and even less proteolipid protein, than do normal adult mice. The "lower layer" fraction has only small amounts of basic and proteolipid protein (possibly accounted for by the material identifiable as myelin in Fig. 3). Agrawal et al. (1) have also observed that the protein composition of their "myelin like" fraction and of mature myelin is similar, except for the absence of basic protein in the "myelin-like" material. Proteolipid protein was not clearly resolved under their electrophoretic conditions. The pellet fraction from Quaking mice has only trace amounts of basic and proteolipid protein.

Although the disorder in Quaking can clearly be broadly characterized as an arrest in formation of myelin ("hypomyelinogenesis") we have no information as to what the genetic lesion might be. It might possibly be a lesion involved in synthesis of a myelin specific lipid, e.g. chain elongation to form long chain fatty acids, or galactosylation of ceramide to form cerebroside. Our data, demonstrating an analogy between the myelin protein of Quaking mouse and of immature controls, opens the possibility that the lesion might directly effect synthesis of one of the myelin proteins. A more interesting interpretation of the data is that it indicates a coordinate control mechanism governing myelinogenesis, since both lipid and protein components of myelin are affected by what we assume to be a single genetic lesion. If this is the case then the direct effect of the Quaking gene need not be at the level of synthesis of a structural element of myelin or even involve oligodendroglial cells; e.g. the initial expression of the lesion might be at the level of recognition between neuronal axons and oligodendroglial cells.

It is possible to relate our observations concerning "lower layer" material to the hypomyelinogenesis in Quaking mice. A simple interpretation is dependent on a hypothesis, suggested initially by Agrawal et al. (1), that the "myelin like" or "lower layer" material is involved in myelinogenesis, either as a precursor or as a framework for assembly of mature myelin. The only direct evidence to support this hypothesis is the similarity of protein composition detailed above. The ultrastructural characterization of lower layer material as a membranous fraction is compatible with this suggestion but offers no independent support. Since the "lower layer" material does not accumulate in Quaking mice we propose that the genetic lesion is expressed at the time when the myelin like material is being formed, presumably as a result of modification of oligodendroglial cell membrane, and not at the stage of final assembly of mature myelin. In a final flight of speculation, it is possible that the pellet fraction accumulating in Quaking mice is precursor to the "lower layer" or "myelin like" material.

ACKNOWLEDGEMENTS

This research was supported by Public Health Service Grants NS-09094, NS-02476, NS-03356 and MH-06418.

REFERENCES

1 AGRAWAL, H.C., BANIK, N.L., BONE, A.H., DAVISON, A.N., MITCHELL, R.F. and SPOHN, M., Biochem. J., 120 (1970) 635.
2 BANIK, N.L. and DAVISON, A.N., Biochem. J., 115 (1969) 1051.
3 BAUMANN, N.A., JACQUE, C.M., POLLET, S.A. and HARPIN, M.L., Europ. J. Biochem., 4 (1968) 340.
4 BERGER, B., Brain Res., 25 (1971) 35.
5 BOWEN, D.M. and RADIN, N.S., J. Neurochem., 16 (1968) 457.

6 COSTANTINO-CECCARUBU, E. and MORELL, P., Brain Res., 29 (1971) 75.

7 GONZALEZ-SASTRE, F., J. Neurochem., 17 (1970) 1049.

8 GREENFIELD, S., NORTON, W.T. and MORELL, P., J. Neurochem., 18 (1971) 2119.

9 HOGAN, E.L. and JOSEPH, K.C., J. Neurochem., 17 (1970) 1209.

10 LAEMMLI, U.K., Nature (Lond.), 227 (1970) 680.

11 MARTENSON, R.E., DIEBLER, G.E. and KIES, M.W., J. Biol. Chem., 244 (1969) 4268.

12 NESKOVIC, N., NUSSBAUM, J.L. and MANDEL, P., Brain Res., 21 (1970) 39.

13 NORTON, W.T., in Chemistry and Brain Development (Ed. PAOLETTI, R. and DAVISON, A.) Plenum Press, New York, 1971, p.327.

14 RAINE, C.S., PODUSLO, S.E., and NORTON, W.T., Brain Res., 27 (1970) 11.

15 REASOR, M.J. and KANFER, J.N., Life Sci., 8 (1969) 1055.

16 SIDMAN, R.L., DICKIE, M.M. and APPEL, S.H., Science, 144 (1964) 309.

17 SINGH, H., SPRITZ, N. and GEYER, B., J. Lipid Res., 12 (1971) 473.

18 WISNIEWSKI, H. and MORELL, P., Brain Res., 29 (1971) 63.

19 WOLFGRAM, F. and KOTORII, K., J. Neurochem., 15 (1968) 1281.

MYELIN ENZYMES AND PROTEIN METABOLISM

Neville MARKS

New York State Institute for Neurochemistry and Drug Addiction- Ward's Island, New York 10035 (USA)

Turnover of myelin has attracted interest in relation to clinical and experimental diseases involving demyelination (7,11). In adult rats, myelin represents about 12 % of total fresh weight, of which some 25 % is protein (14) and earlier concepts for its relative inertness are difficult to reconcile with the known half-life of under 20 days for 90 % of brain proteins (11).Turnover encompasses both synthesis and breakdown, linked together in a manner to preserve the function and integrity of tissues. Breakdown,like synthesis, is an orderly process involving a large number of hydrolases (proteinase and peptide hydrolases) acting in a sequential manner for the production of amino acids and peptides (Fig. 1). Earlier studies from this laboratory have shown the presence of proteinases in crude fractions containing myelin (7, 10) but there has been some question concerning the degree of contamination by enzymes from other particulates. All studies on turnover in myelin must consider 1) the purity and structural nature of isolated material, 2) the relative metabolism of different layers of the myelin sheath, 3) the role of the cytoplasmic inclusions or other discontinuities within the myelin sheath, 4) in studies on incorporation, questions related to penetrability of the precursor, 5) the role of Schwann or satellite cells, and 6) the age of the animal.

In this study, protein metabolism (breakdown and synthesis) in crude myelin fractions (M_1) was compared to material that was water shocked to remove cytoplasmic contamination (M_2), and with the purified membranes (M_3). Adult male rat (35-40 days) or pig brain was homogenized in 10 volumes(w/v) of 0.32 M sucrose and centrifuged to yield the crude mitochondrial fractions (P_2) which contain synaptosomes and myelin, by methods previously described (9).

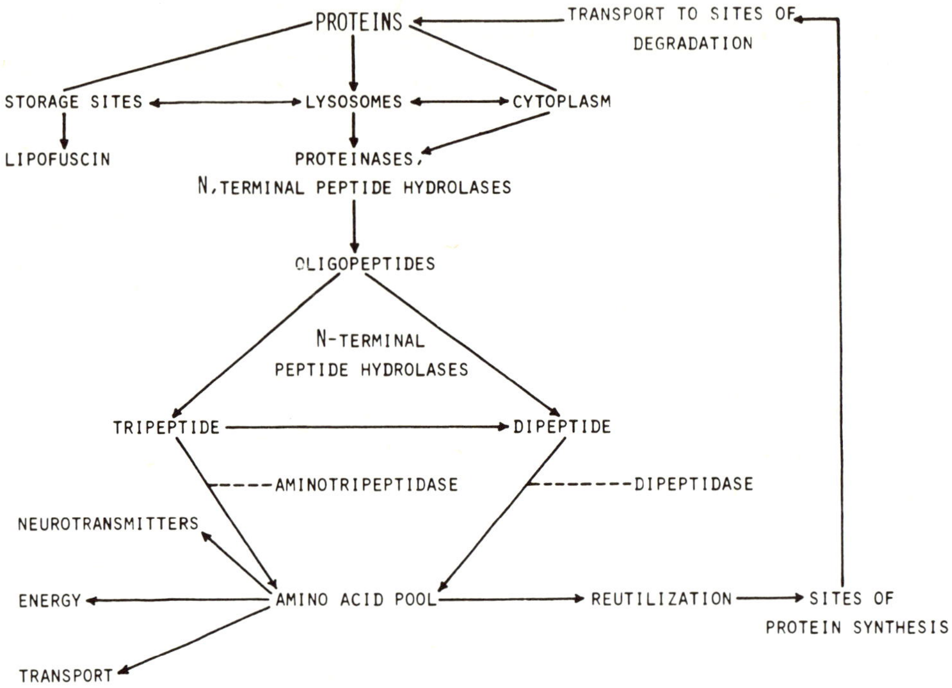

Fig. 1. Postulated scheme for protein turnover. Although some mechanisms of synthesis are known, knowledge of breakdown processes and the sequence of reactions for degradation of protein are unknown. The above scheme is based on the peptide specificities of purified brain enzymes (11).

Crude myelin (M_1) was separated from other particulates by centrifugation of the P_2 fraction on a continuous sucrose gradient, 0.8-1.4 M, using a Spinco SW 25 rotor (24,000 rev./min for 1 h) (9). Following centrifugation, the material in each tube was divided into 17 fractions : fractions 13-15 (interface 0.32 - 0.8 M) represent enriched myelin, and the pellet sedimented at 1.4 M sucrose represents purified (heavy) mitochondria. The myelin-rich fractions were pooled, centrifuged at 50,000 g for 1 h, and then treated with 20 volumes of cold-distilled water for 1 h at 0° and recentrifuged to yield the water-treated pellet M_2. This material was purified further by suspension in 5 ml of 0.3 ml CsCl and then placed over a discontinuous gradient of CsCl consisting of 5 ml quantities of 0.5, 0.7, 0.9 and 1.1 M CsCl in a 1 x 3 inch cellulose tube, and centrifuged in a Spinco SW 25 rotor at 24,000 rev./min for 1 h. The location of fractions M_3 (purified membranes) and M_4 (fragmented membranes and vesicle-like structures) is illustrated in Fig.2. It is possible that the M_4 fraction has some similarities to the "premyelin" material first described by Davison and his group (1).

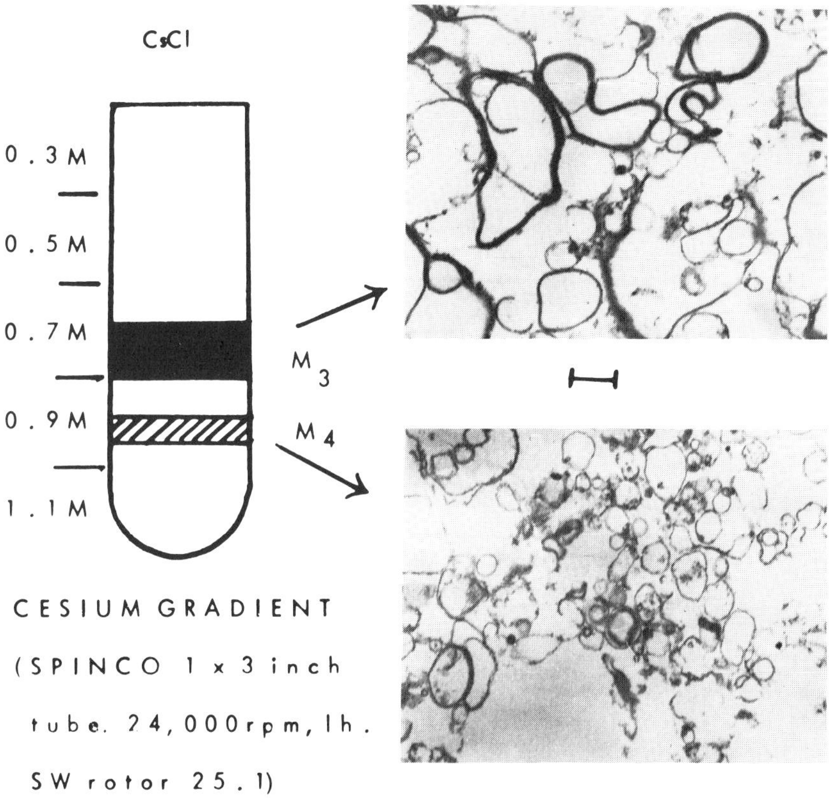

Fig. 2. Method for the purification of the water washed pellet (M_2) to yield the CsCl fractions M_3 and M_4. Electronmicrographs fixed with OsO_4. Note the appearance of fragmented membranes and vesicles in the lower fraction.

Myelin (M_3) was fragmented into its protein and proteolipid moieties by the Triton X-100 : acetate procedure of Eng et al. (5). As noted by Norton in his review (14) there is a notable lack of information on the nature of most myelin proteins and the manner they are complexed with lipids (i.e., proteolipids). The Eng procedure with slight modifications (5) can yield fractions consistent with the Wolfgram, Folch-Lees, and basic protein components. Disc-gel electrophoresis in phenol-urea solvents revealed some cross contamination (3). Quantitative methods for isolation of proteolipids have attracted considerable interest, as noted by others at this symposium (12, 13). There are still many problems associated with identification and separation, and other methods of promise must be carefully evaluated for use in studying turnover.

In the present study, adult rats received intracisternal doses of ^{14}C-lysine at 0.25 - 6 h prior to sacrifice. Comparison of

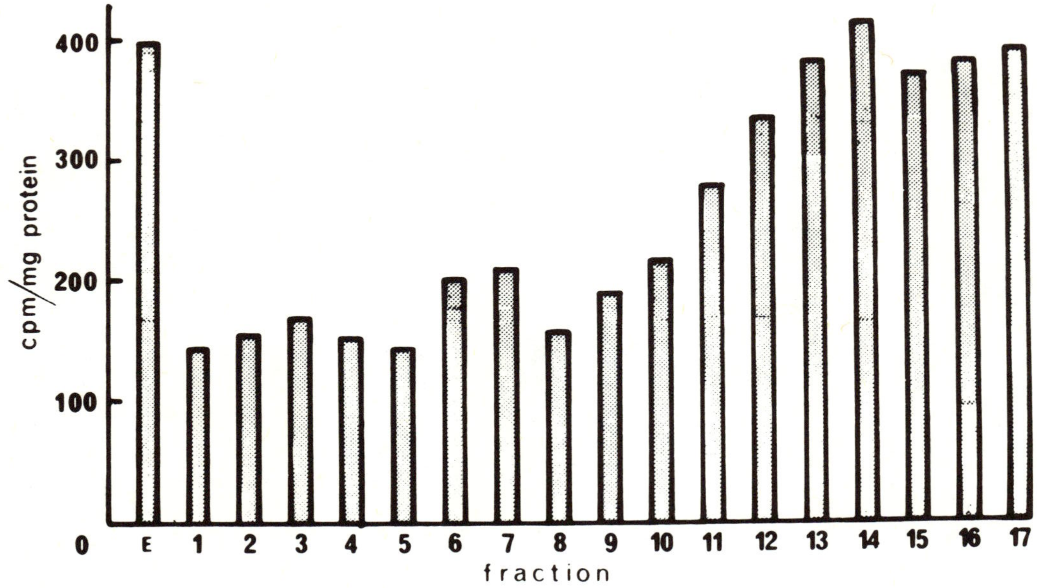

Fig. 3. Incorporation of ^{14}C-lysine into subfractions of crude mitochondria containing myelin (in 0.32 M sucrose) after separation on a continuous sucrose gradient (0.8 - 1.4 M). Fraction E represents purified (heavy) mitochondria sedimenting at 1.4 M and fractions 12-16 the enriched myelin regions (0.32 - 0.8 M) (Fraction M_1). Adult rats (35 days of age) received 1 μCi of label intracisternally 40 min prior to sacrifice and preparation of the crude mitochondrial fractions. Results are the means of four determinations agreeing within 15 %. Note the high incorporation into the enriched myelin fractions comparable to that in the purified mitochondrial pellet. The intermediate fractions contain a variety of particulates including nerve-endings, lysosomes and small mitochondria (11).

incorporation into different crude mitochondrial sub-fractions showed a surprisingly high incorporation into the myelin-rich regions (Fig. 3). At first, this was attributed to contamination, since crude fractions are known to contain entrapped particulates (2). Despite the many washing and purification procedures, the M_2 and M_3 fractions still displayed considerable radioactivity. Many early studies on protein turnover failed to take into account the many factors that affect measurement of half-life in single-pulse type experiments (11). These include variables such as transport factors (route of injection, and the extent to which the precursor pool is labelled), metabolic conversion of the precursor, reutilization of amino acids formed by breakdown etc. (7). To overcome some of these difficulties, $t_{1/2}$ was measured on the basis of the time taken to replace 50 % of the known lysine content of the different fractions (directly determined after 6 N HCl hydrolysis on an autoanalyzer). Actual lysine residues incorporated into proteins were conveniently estimated by calculating the radioactivity representing 1 nmole in the total pool of free lysine (known to be approximately 0.3 µmole per g wet weight in rat cerebrum). Radioactivity in the pool itself at different pulse periods was measured in aliquots of the perchloric acid-soluble fraction of whole rat brain homogenate plus washings from the interior of the skull (Table I). Based on the specific activity of the protein fractions (counts/min/mg protein)

TABLE I

Change in Specific Activity on the Free Pool with Time Following Intracisternal Injection of ^{14}C-Lysine

Min.	Specific activity of free pool a	b
15	628	628
40	355	330
90	161	193
180	100	98
360	54	43

a, counts representing one nmole of lysine following injection of 1 µCi (based on concentration in the pool of free amino acids of 0.3 µmole/g fresh weight). b, values based on the change between consecutive time points.

the number of µmoles ^{14}C-lysine incorporated was calculated and expressed as a rate (r) (per cent conversion per hour of the known lysine content of protein). An expression for $t_{1/2}$ can be derived from : $r = 100\ (1 - \delta)/t$ where δ is the fraction converted, i.e., $Ct = Co\ (1 - \delta)$ where Co is the initial concentration (100) and Ct that at time t. In the special case of 50 % conversion (Co = 50), $t_{1/2} = \frac{\ell\ 2}{k}$ where k is the rate constant. For very low rates of conversion $(\delta \ll 1), \ell n(1-\delta) = -Kt\ \ \Omega\ \ -\delta$. Substitution in these expressions in the equation for first order yields $t_{1/2} = \frac{t\ell n2}{24r}$ or $\frac{2.8}{r}$ (days).

Results gave evidence of both short and long-lived proteins. The highest incorporation was observed in the basic proteins (fraction III) with $t_{1/2}$ of 14-21 days, followed by the Folch-Lees material 12-14, and the insoluble material (fraction I) 28-32 days (Table 2). Incorporation at very short pulse periods up to 40 min

TABLE II

Half-Lives ($t_{1/2}$) of Myelin Proteins of Rat Brain Following Intracisternal Injection of ^{14}C-Lysine

Pulse period (min)	Myelin	
	% Incorporation	$t_{1/2}$ (days)
Fraction I		
40	0.10	29
90	0.12	24
180	0.11	26
360	0.10	28
Fraction II		
40	0.30	10
90	0.11	26
180	0.10	29
360	0.12	24
Fraction III		
40	0.25	12
90	0.27	11
180	0.16	18
360	0.14	21

Per cent incorporation and the half-lives ($t_{1/2}$) were calculated from the simplified equation for first order kinetics. Following intracisternal injection of ^{14}C-lysine (1 µCi) to rats, the specific activity of the precursor pool was determined (Table 1) and from this data nmole lysine between consecutive time points calculated and expressed as per cent incorporation of proteins with known lysine composition/hour.

TABLE III

Alteration in Hydrolases with Purification of Rat Brain Myelin Fractions

	Acid proteinase		Neutral proteinase		Aminopeptidase Leu-Gly-Gly		2',3'-AMP phosphatase	
	Act.	S.A.	Act.	S.A.	Act.	S.A.	Act.	S.A.
Myelin M_1	(1.9)+	93	(0.6)	28	(4.3)	21	(3.8)	18
M_2	19	40	43	26	51	24	20	22
M_3	7	52	0	0	20	32	20	28

+Activity in parentheses represents μmole /g wet weight/hour and other values the percentage recovery based on that in M_1. Specific activity expressed as μmole /g protein/hour. Note the increase in specific activity with purification in the case of aminopeptidase and the phosphatase with a 20 percent recovery.

M_1 represents crude myelin, M_2 the water washed fraction and M_3 the myelin membrane fraction after purification on CsCl gradients.

TABLE IV

Arylamidases in Purified Fractions of Myelin

	Monoacyl- Arg-βNA		Dipeptidyl- Arg-Arg-		Dipeptidyl- Lys-Ala-βNA	
	Act.	S.A.	Act.	S.A.	Act.	S.A.
M_1	(11)†	53	(3.3)†	16	1.8	5
M_2	12	14	12	4	26	5
M_3	5	20	4	5	5	4

†Activity in parentheses represents μmole/g wet weight/hour and other values the percentage recovery based on that of M_1. Specific activity expressed as μmole/g protein/hour. Note fall in specific activity and recovery with purification. See Fig. 3 for identification of fractions.

was slower than the comparable mitochondrial fractions (2, 9), probably as a result of the penetration of these structures by label. Short pulse periods were used to reduce the corrections required for reutilization of amino acids formed by breakdown of labelled proteins. Lysine was selected as precursor since it is subject to less conversion, is present in low concentration in the free pools, and is present in relatively high concentration in basic proteins.

Breakdown, the other component of turnover, was evaluated by measurement of proteinases and peptidases by the methods previously published (7, 8). Proteinase activity was measured in the absence (endogenous) and presence of substrate (denatured hemoglobin) at acid (3.2) and physiological (7.6) pH : N-terminal peptidases were measured with the following substrates : Leu-Gly-Gly, monoacyl, and dipeptidyl arylamides. These hydrolases were compared to the known myelin marker, ribonucleoside 2',3'-cyclic phosphate esterase, measured by the method of Olafson _et al._ (15). Purification of crude myelin was accompanied by a loss of most hydrolases studied except the aminopeptidase (Leu-Gly-Gly) and the cyclic phosphatase (Table 3). Removal of cytoplasmic contamination led to a 57 percent reduction of neutral proteinase, and a large fall in acid proteinase (Table 3). Further purification on sucrose and CsCl gradients led to a total loss of neutral proteinase and only trace activities of acid proteinase. In the case of arylamide substrates, 80 percent or more of the activity was lost on water treatment (M_2) and 95 percent or more after purification on CsCl (Table 4). Cs^+ up to 1 M was itself without effect in all assay procedures.

Conclusions

Present findings together with those of other groups (11, 18 - 21) challenge the established view that myelin is metabolically inert with respect to its protein or proteolipid components. Discrepancies between findings for turnover can arise from inadequate recognition of the many factors affecting their measurement (as already noted). Age is an important factor since the slow deposition of myelin can continue well beyond the normal period of active growth. Sammeck and Brady (18) have demonstrated considerable incorporation into myelin basic proteins, and Wood and King (21) have shown half-lives for the same components in young and old rats of 21 days ; M. Smith and co-workers (19) have observed half-lives for proteolipids _in vitro_ and recently _in vivo_ (20) of under 40 days. Studies on protein turnover in brain at all ages indicate a diversity of half-lives with an average of about 15-20 days, although minor components show a range extending from hours to several weeks. In no case does the percentage of "stable" components account for the relatively large amounts of myelin present in the adult nervous system. Largely from such considerations it seems unlikely that myelin is inert, although its turnover may become progressively slower with age. It might be of some interest to carefully compare turnover in adults of different ages, particularly that of young adults with that of senile animals. In many ways the concept of myelin "inertness" is an attractive hypothesis since it could account for some aspects of nerve function (efficiency of nerve conduction, homeostatic mechanisms, etc.). Pathological changes, as is well known, lead to severe malfunction of the nervous system.

The elimination of proteinases and arylamidases with purification suggests that these enzymes are not true components of the membrane fractions represented by M_3. Activity in crude fractions is more difficult to interpret. Even in the water-washed fractions (M_2),activity of most hydrolases was 40-50 percent of that in the crude M_1 material. Such data might indicate the presence of hydrolases in soluble or other structures easily removed in purification on sucrose or CsCl gradients. The sheath is known to contain cytoplasmic inclusions (Schmitt-Lantermann clefts) and these could account for breakdown observed in the sheath during pathological conditions. We have postulated that release of acid proteinases could lead to the degradation of basic proteins with formation of encephalitogenic peptides (4,6,7). The source of proteinase and mechanisms by which it releases peptides from myelin are unknown. Studies with purified acid proteinase and purified basic protein show rapid breakdown. The nature of the susceptible bonds of basic protein of myelin is consistent with the known specificity of purified acid proteinase of brain (7). The present results are at variance with those of Riekkinen and Clausen (16,17) in that they observed considerable neutral proteinase even after extensive purification of myelin on a series of sucrose gradients. Assay of neutral proteinase is subject to considerable difficulties, and some even have challenged its existence in tissues (see ref. 11). Activities reported by Riekkinen and Clausen were of low order, since up to 24 hours incubation was required to detect activities.

The heterogeneous rates of turnover of the various components of the myelin sheath pose some intriguing questions about the mechanisms involved, and point out the possible alterations this structure may undergo. Highly purified myelin lamellae, themselves, clearly do not contain the appropriate machinery for turnover, implying that other components of the sheath or adjacent structures supply the essential factors for synthesis and breakdown. Various structures within the myelin sheath could act as reservoirs for cytoplasmic factors necessary for synthesis ; these include the Schmitt-Lantermann clefts in the CNS, and nodes of Renvier in the PNS. Penetration of precursors between layers of the sheath is a distinct possibility based on recent findings. As in the case of other membranes, the integrity of structure is dependent on the integration of the mechanisms controlling turnover of both lipid and protein moieties. Alteration in these processes could account for the etiology of many diseases of clinical importance.

REFERENCES

1 AGRAWAL, H.D., BANIK, N.L., BONE, A.H., DAVISON, A.N., MITCHELL, R.F.and SPOHN, M., Biochem. J., 120 (1970) 635.

2 D'MONTE, B., MARKS, N., DATTA, R.K. and LAJTHA, A., in Symposium on Protein Metabolism in the Nervous System, Plenum Press, New York, 1970, p.185.

3 D'MONTE, B., MELA, P. and MARKS, N., Europ. J. Biochem., 23 (1971) 355.

4 EINSTEIN, E.R., CSEJTEY, J. and MARKS, N., FEBS Letters, 1 (1968) 191.

5 ENG, L.F., CHAO, P.C., GERSTL, B., PRATT, D. and TAVASTSTJEMA, M.G., Biochemistry, 7 (1968) 4455.

6 LAJTHA, A. and MARKS, N., in Dynamics of Protein Metabolism in the Nervous System (Ed. BOGOCH,S.) Plenum Press, New York,1969, p.181.

7 LAJTHA, A. and MARKS, N., in Handbook for Neurochemistry (Ed. LAJTHA, A.) Plenum Press, 5B, New York, 1971, p.551.

8 MARKS, N., Int. Rev. Neurobiol., 11 (1968) 57.

9 MARKS, N., D'MONTE, B., BELLMAN, C. and LAJTHA, A., Brain Res., 18 (1970) 309.

10 MARKS, N. and LAJTHA, A., Biochem. J., 89 (1963) 438.

11 MARKS, N. and LAJTHA, A., in Handbook for Neurochemistry (Ed., A. LAJTHA) Plenum Press, 5A, New York, 1971, p.49.

12 MEHL, E.

13 MORELL, P.

14 NORTON, W.T., in Cellular and Molecular Basis of Neurology (Ed. GOLDSTEIN, E. and APPEL, S.) Lee and Febiger, Philadelphia, in press.

15 OLAFSON, R.W., DRUMMOND, G.I. and LEE, J.F., Canad. J. Biochem., 47 (1969) 961.

16 RIEKKINEN, P.J. and CLAUSEN, J., Brain Res., 15 (1969) 413.

17 RIEKKINEN, P.J. and CLAUSEN, J., Brain Res., 19 (1970) 213.

18 SAMMECK, R. and BRADY, R.O., in Proc. Intern. Soc. Neurochem., Akademiai Kiado, Budapest, 1971, p.78.- Brain Res., 34 (1971) 241.

19 SMITH, M.E. and HASINOFF, C.M., J. Neurochem., 18 (1971) 739.

20 SMITH, M.E. and RAWLINS, F.A., in Proc. Intern. Soc. Neurochem., Akademiai Kiado, Budapest, 1971, p.77.

21 WOOD, J.G. and KING, N., Nature, 229 (1971) 56.

CONCLUDING REMARKS

A.N. DAVISON

Department of Neurochemistry, Institute of Neurology
Queen Square, London WC 1N 3BG (England)

In concluding this symposium I should pick out a few interesting points which have taken my attention. I think the report by Dr. Folch of bound fatty acids attached covalently to the proteolipid apoprotein is a most important observation. This gives us an important clue to the special physico-chemical properties of the proteolipids and we shall be hearing a lot more about the chemistry of these unique proteins.

A number of papers have raised the question of the purity and identity of myelin. Is the preparation obtained by repeated washing of isolated myelin the true myelin complex, or is it simply the membrane denuded of all water extractable material? Of interest here is the report by Dr. Mehl that the Wolfgram protein is a contaminant and not an integral myelin component. The same problem appears in assessing the nature of the myelin in the inborn errors - Jimpy and Quaking mice described by Dr. Morrell and Professor Mandel. A question naturally interesting myself is whether or not, as has been suggested, this material is similar to the myelin-like fraction isolated from developing brain (1, 2).

Further work should clarify this question and establish if the premyelin fraction is glial plasma membrane of if it is smooth endoplasmic reticulum - we may expect to see considerable advance in this field with the availability of methods of separation of glial cells and their plasma membrane.

Another question discussed by Dr. Marks is that of turnover of myelin proteins. It is necessary for much more research to be concentrated in this area to establish if any of the proteins serve as a permanent skeleton for the more labile myelin constituents.

Finally establishment of the nature and biological activity of the basic encephalitogenic protein has been a fascinating story to which a number of groups made important contributions (Kies, Einstein, Carnegie, Lumsden, Kibler, etc.) and the latest work on the primary structure was ably summarized by Dr. Eylar. Dr. Kies told us about the distribution of myelin basic proteins in different species - another line of work contributing to our understanding of the biological action of their proteins. Research on the encephalitogenic protein has reached an exciting phase and considerable advance in this field can be expected in the next few years particularly if the work can be applied to clinical problems.

REFERENCES

1 AGRAWAL, H.C., BANIK, N.L., BONE, A.H., DAVISON, A.N., MITCHELL, R.F. and SPOHN, M., Biochem. J., 120 (1970) 635.

2 BANIK, N.L. and DAVISON, A.N., Biochem. J., 115 (1969) 1051.

SUBJECT INDEX